S0-CAS-663

Cycloaddition Reactions in Carbohydrate Chemistry

ACS SYMPOSIUM SERIES **494**

Cycloaddition Reactions in Carbohydrate Chemistry

Robert M. Giuliano, EDITOR
Villanova University

Developed from a symposium sponsored
by the Division of Carbohydrate Chemistry
at the 200th National Meeting
of the American Chemical Society,
Washington, D.C.,
August 26–31, 1990

American Chemical Society, Washington, DC 1992

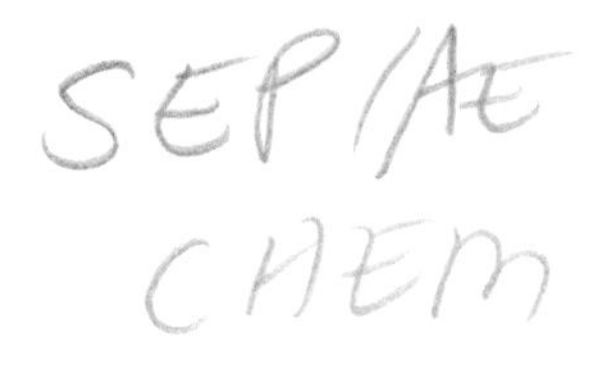

Library of Congress Cataloging-in-Publication Data

Cycloaddition reactions in carbohydrate chemistry /
Robert M. Giuliano, editor

p. cm.—(ACS Symposium Series, 0097–6156; 494).

"Developed from a symposium sponsored by the Division of Carbohydrate Chemistry at the 200th National Meeting of the American Chemical Society, Washington, D.C., August 26–31, 1990."

Includes bibliographical references and index.

ISBN 0–8412–2429–3

1. Carbohydrates—Congresses. 2. Ring formation (Chemistry)—Congresses.

I. Giuliano, Robert M., 1954– . II. American Chemical Society. Division of Carbohydrate Chemistry. III. American Chemical Society. Meeting (200th: 1990: Washington, D.C.). IV. Series.

QD320.C93 1992
547.7'804593—dc20

92–13140
CIP

The paper used in this publication meets the minimum requirements of American National Standard for Information Sciences—Permanence of Paper for Printed Library Materials, ANSI Z39.48–1984. ∞

PRINTED IN THE UNITED STATES OF AMERICA

ACS Symposium Series

M. Joan Comstock, *Series Editor*

Foreword

THE ACS SYMPOSIUM SERIES was founded in 1974 to provide a medium for publishing symposia quickly in book form. The format of the Series parallels that of the continuing ADVANCES IN CHEMISTRY SERIES except that, in order to save time, the papers are not typeset, but are reproduced as they are submitted by the authors in camera-ready form. Papers are reviewed under the supervision of the editors with the assistance of the Advisory Board and are selected to maintain the integrity of the symposia. Both reviews and reports of research are acceptable, because symposia may embrace both types of presentation. However, verbatim reproductions of previously published papers are not accepted.

Contents

INDEXES

Preface

CYCLOADDITION REACTIONS ARE AMONG the most versatile ring-forming reactions in organic chemistry. The Diels–Alder reaction, its heteroatom analogs, and dipolar addition reactions constitute a large part of the available synthetic methodology for the synthesis of both carbocyclic and heterocyclic rings. As in the case of many other reactions used in synthesis, interest in the use of carbohydrate substrates as a means of obtaining enantiomerically pure targets has grown. Carbohydrate-derived dienes and dienophiles, 1,3-dipoles and dipolarophiles, as well as carbohydrate-based chiral auxiliaries have been applied to the synthesis of optically active organic compounds. Volumes 20 and 21 of the Specialist Periodical Reports on Carbohydrate Chemistry, covering the 1986 and 1987 literature, include references to the synthesis of (+)-compactin, (+)-pillaromycinone, diplodiatoxin, octosyl acid A, *C*-nucleosides and carbocyclic nucleosides, and pyrrolizidine alkaloids, all based on the cycloaddition reactions of carbohydrates. In addition to natural products synthesis, cycloaddition reactions of sugar substrates have also been investigated with the aim of understanding the basis for the high levels of stereoselectivity that often are observed.

The purpose of the symposium upon which this book is based was to present to the chemical community the recent advances that have been made in the cycloaddition reaction chemistry of carbohydrates. The symposium included presentations on inter- and intramolecular Diels–Alder reactions, dipolar addition reactions, the use of carbohydrate-derived chiral auxiliaries, and theoretical studies. The synthesis of carbohydrate substrates for cycloaddition reactions was highlighted throughout the meeting. Prior symposia on the cycloaddition reactions of carbohydrates had not been held, nor had any review articles been published on this subject. The successful response to the symposium and the absence of a review on the topic prompted the undertaking of this volume. The focus of the book is on the use of carbohydrate-derived substrates in cycloaddition reactions. It is hoped that the information presented will be of value to those interested in the organic chemistry of carbohydrates, and that further investigations of cycloaddition reactions of carbohydrates will continue to reveal novel and useful synthetic methodology.

I wish to thank the authors of the chapters in this volume and their co-workers, whose efforts in the study of carbohydrate cycloaddition reactions provided the impetus for organizing a symposium and a book. I also wish to express my appreciation for the generous support received for the symposium from the ACS Division of Carbohydrate Chemistry, the General Electric Company, Glaxo Research Laboratories, the Petroleum Research Fund, Rhône-Poulenc-Rorer Central Research, Schering-Plough Corporation, and Wyeth-Ayerst Research.

ROBERT M. GIULIANO
Villanova University
Villanova, PA 19085

February 1992

Chapter 1

Cycloaddition Reactions in Carbohydrate Chemistry

An Overview

Robert M. Giuliano

Department of Chemistry, Villanova University, Villanova, PA 19085

A review of cycloaddition reactions in carbohydrate chemistry is presented. The use of carbohydrate-derived dienes and dienophiles in the Diels-Alder reaction, hetero-Diels-Alder and dipolar addition reactions of carbohydrates are described. Stereochemical aspects of the cycloaddition processes are also discussed, and applications to the synthesis of natural products are included for each reaction type. Much of the material presented has appeared in the literature within the past five years; however, earlier studies are also included in order to give a more representative historical perspective.

Diels-Alder Reactions

Cycloadditions of conjugated dienes to alkenes (dienophiles) provide rapid access to substituted cyclohexenes. In view of the widespread occurrence of carbocyclic compounds in nature, and the recent demands for the synthesis of enantiomerically pure compounds, it is not surprising that carbohydrates were examined as substrates for Diels-Alder reactions. Both carbohydrate-derived dienophiles and dienes have been described in the literature. The products of reactions involving these types of compounds, referred to as annulated sugars, are most often formed stereoselectively. The high levels of both diastereofacial selectivity and endo/exo selectivity sometimes result in the formation of a single cycloadduct from among several possibilities.

0097–6156/92/0494–0001$06.75/0

Carbohydrate Dienophiles. The most popular carbohydrate dienophiles to date have been pyranoid enones. Examples of Diels-Alder reactions involving carbohydrate 2,3-enones, 3,4-enones, and enolone esters are described below. In landmark studies reported by Fraser-Reid and co-workers, carbohydrate-derived 2,3-enones were shown to react with dienes to give cyclohexopyranosides (*1*- *3*). Derivatives of methyl and ethyl 2,3-dideoxy- α **D**-*glycero*-hex-2-enopyranosid-4-ulose **1** underwent cycloaddition with butadiene in the presence of aluminum chloride, and with oxygen -substituted dienes in the absence of any catalyst. Annulated pyranosides **2** and **3** were obtained

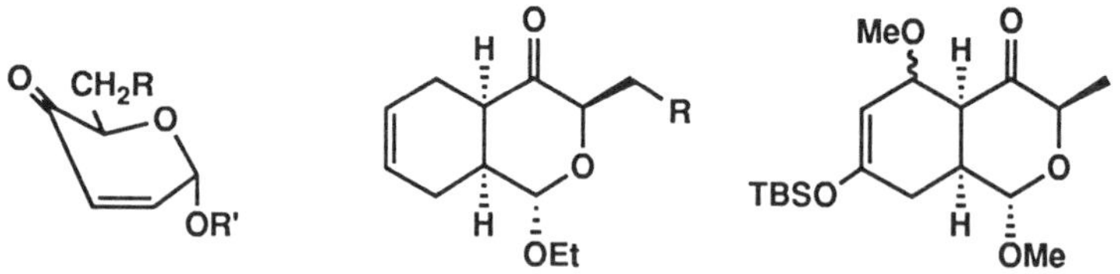

1 2 3

R'=Me,Et

R=H,OAc,OTr,OTBS

stereoselectively in high yield, with cycloaddition occurring from the face of the enone opposite the anomeric substituent. Further transformations of these products along synthetically useful pathways were explored in Fraser-Reid's laboratories. An example is the synthesis of the antibiotic actinobolin **7**, outlined in Scheme 1 (*4*). The amino group was introduced into **3** by conversion to the oxime and lithium aluminum hydride reduction. Oxidation with lead tetraacetate gave **5**. By taking advantage of the conformational properties of subsequent intermediates in the synthesis, the authors were able to introduce the diol functionality in **7** with stereocontrol. An interesting question arose out of the actinobolin synthesis concerning the stereochemistry of the cycloaddition step. One might assume that the facial selectivity of the Diels-Alder reaction would be determined by the anomeric substituent; however, it was found that a single cycloadduct β-endo was obtained from *either* **8** or **9** when treated with Danishefsky's diene. The pyranose ring oxygen is believed to affect both the reactivity and the stereoselectivity of these enones relative to their carbocyclic counterparts (*5*). The effect of the ring oxygen is ascribed to a lowering of the dienophile LUMO energy. Effects of other oxygen

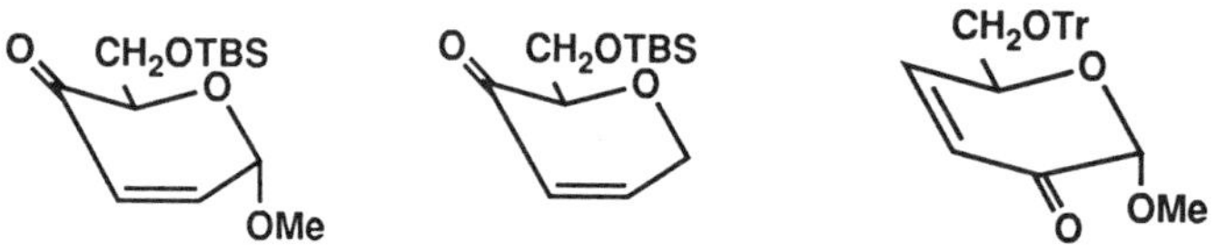

8 9 10

substituents on the pyranose ring were also observed. For example, it was found that the presence of an exocyclic oxygen at the α'-position, as in **10**, dramatically lowered the dienophilicity.

The effect of high pressure on cycloaddition reactions involving carbohydrates as substrates has been extensively investigated by Jurczak and coworkers (*6-8*). In their studies of Diels-Alder reactions of pyranoid enones, improvements in stereoselectivity were observed when the reactions were carried out at high pressure (*6*).

11 kbar
ether, 20 h
91%

diastereoselectivity

high pressure: 98:2
thermal : 80:20

In studies by Card, Diels-Alder reaction of carbohydrate enones with 1-[(trimethylsilyl) oxy]-1,3-butadiene, followed by treatment of the products with DDQ in refluxing benzene proved to be an efficient route to benzannelated pyranosides (*9*).

DDQ
Δ

11
50-70%

Glycals have been utilized as dienophiles in a Diels-Alder-based route to an aglycone of the aureolic acid antibiotics (*10*). Franck and co-workers investigated the cycloaddition of a quinone methide derived from cyanobenzocyclobutene **12** with glycal **13** in their early studies of the synthesis of a model aglycone for olivin (Scheme 2). The stereochemistry of the product **14** is that which results from the addition of the diene to the face of **13** opposite the methoxy group. Addition occurs in both endo and exo modes. Subsequent transformations leading from **14** to the target were highly efficient, and the target tetralone **15** was obtained in 21% overall yield. In more recent studies of

Scheme 1

Synthesis of Actinobolin

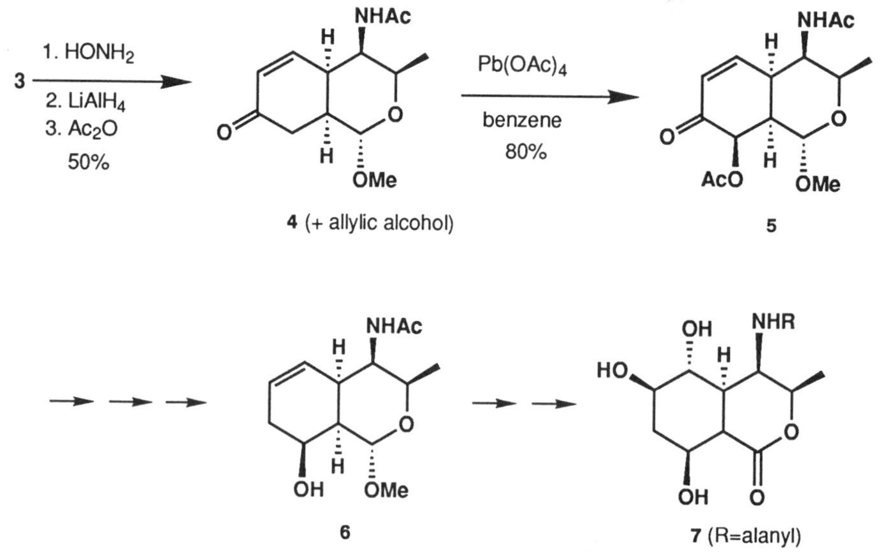

3 → 1. $HONH_2$ 2. $LiAlH_4$ 3. Ac_2O 50% → 4 (+ allylic alcohol) → $Pb(OAc)_4$ benzene 80% → 5

→ → → 6 → → 7 (R=alanyl)

Scheme 2

Synthesis of a Model Aglycone for Olivin

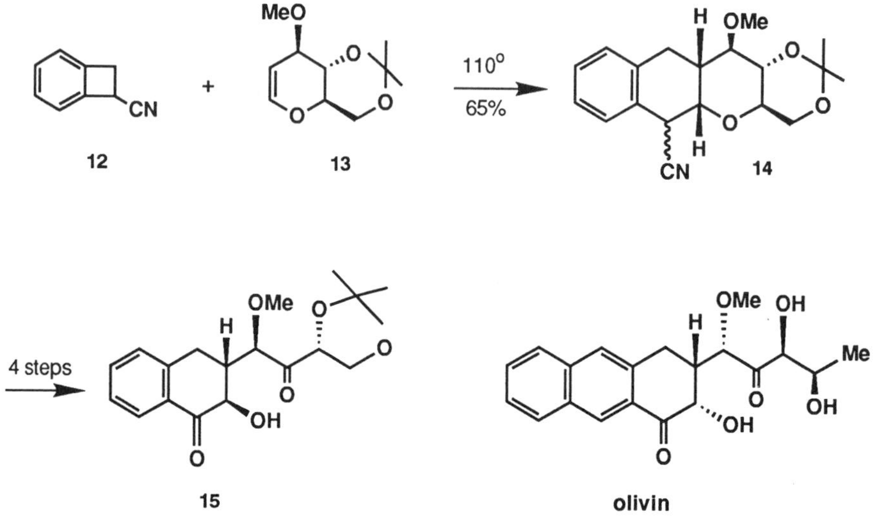

12 + 13 → 110° 65% → 14

4 steps → 15

olivin

aureolic acid antibiotic synthesis from the Franck laboratory, enone **17** was reacted with *E*-1-trimethylsilyoxybutadiene to give **18** (*11*).

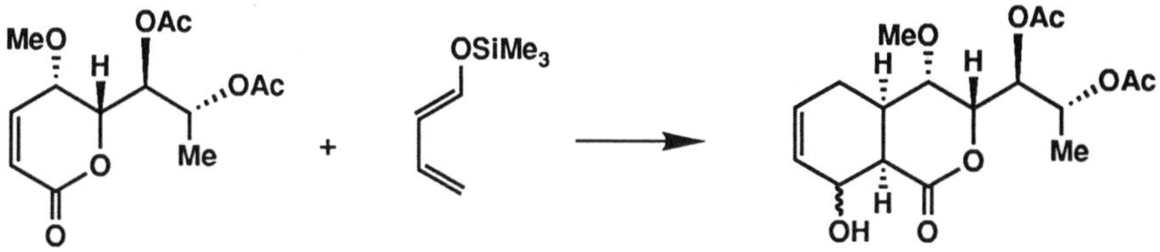

Carbohydrate-derived enolone esters differ from the enones discussed above in that the olefinic bond is substituted with an acyloxy group. The chemistry of enolone esters, in particular, their application to the synthesis of enantiomerically pure non-carbohydrate natural products, has been studied extensively by Lichtenthaler and co-workers (*12*). In a recent report, the Diels-Alder reactions of pyranoid enolone esters with cyclopentadiene was carried out under thermal, Lewis acid catalysis, and high-pressure conditions (*13*). The presence of the acyloxy group on the double bond renders the enolone esters less reactive than the enones which have a hydrogen at this position; however, high yields of cycloadducts were generally obtained in both the high-pressure and Lewis acid catalyzed reactions. For example, enolone ester **19** gave greater than 80% yields of the α-endo cycloadduct when treated with cyclopentadiene in the presence of titanium tetrachloride,

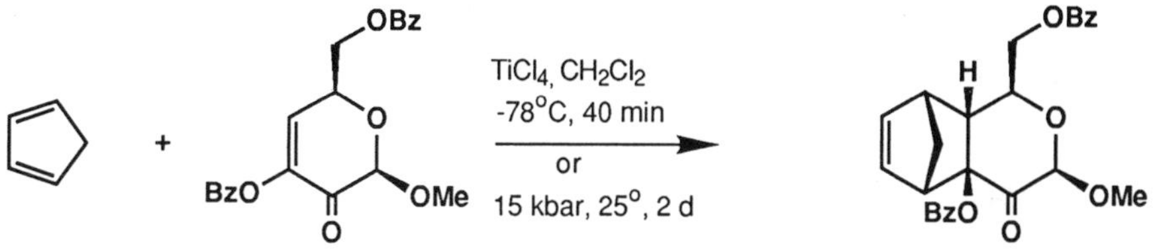

$TiCl_4$, CH_2Cl_2, -78°C, 40 min or 15 kbar, 25°, 2 d

19 → **20** α-endo (>80%) + isomers (12%)

or at 15 kbar. Endo/exo ratios ranged from 5:1 to 9:1 for the four enolone esters that were studied, while diastereofacial selecivities were as high as 40:1. The facial selectivities were ascribed to addition of the diene to the sterically less hindered side of the enolone in its preferred ground state conformation, as determined by molecular mechanics calculations and NMR methods.

Horton and Bhata reported the synthesis of functionalized carbocycles by the Diels-Alder reactions of cyclopentadiene with levoglucosenone (*14*) . The addition of cyclopentadiene occurred predominantly form the face of the enone opposite the 1,6-anhydro bridge, in the endo mode.

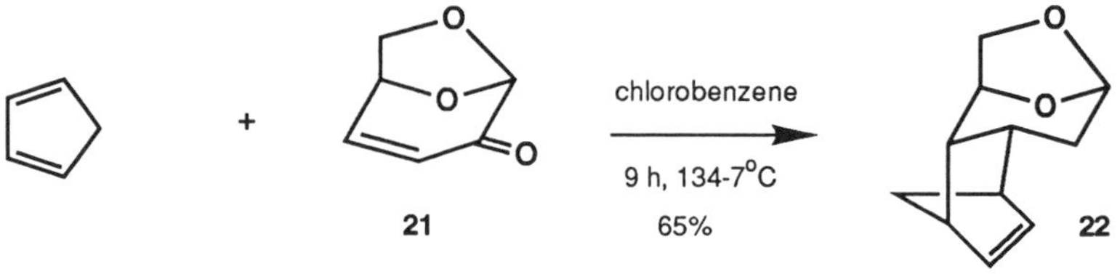

chlorobenzene

9 h, 134-7°C

65%

21 **22**

The carbohydrate dienophiles in the above examples are all derived from cyclic structures. Acyclic sugar dienophiles have also been studied. These substrates are accessible through the reaction of a carbohydrate with a stabilized Wittig reagent or by fragmentation reactions. Two examples of the use of acyclic carbohydrate dienophiles in Diels-Alder reactions are presented below.

Horton and co-workers reported the reaction of cyclopentadiene with α,β–unsaturated ester **23**, derived from **D**-arabinose (Scheme 3). The major product **24** was transformed into a prostaglandin precursor (*15*). Compound **25** has the correct relative stereochemistry of all five stereocenters of PGF_2, and also the same absolute stereochemistry at corresponding positions.

Scheme 3

Synthesis of a Prostaglandin Precursor

CO_2Me OAc AcO AcO CH_2OAc

toluene $AlCl_3$ 57%

MeO_2C OAc AcO AcO CH_2OAc

1. OsO_4- $NaIO_4$ 2. $NaBH_4$ 3. Ac_2O

O O AcO OAc AcO AcO CH_2OAc

23 **24** **25**

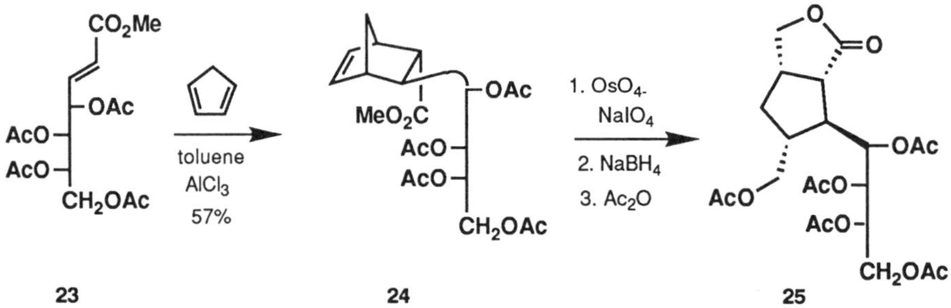

In their work on aureolic acid antibiotic, Franck and co-workers examined the Diels-Alder reaction of the quinone methide derived from **26** with sugar dienophile **27** (Scheme 4). Heating of **26** in a sealed tube for 3 days in chlorobenzene afforded cycloadducts **28** and **29** in a 4:1 ratio in 71% yield. The major product has the *S*-configuration at C-3 and results from the addition of the diene to the *si* face of the dienophile, in the endo mode, as shown below (*16*).

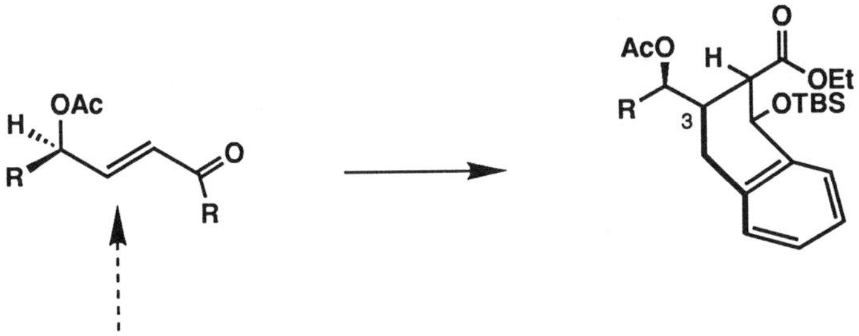

The facial selectivity was thought to arise from a stereoelectronic bias imposed on the transition state by the allylic acyloxy substituent. Addition occurs *anti* to the acyloxy group in the dienophile conformer in which the C-H bond lies approximately in the plane of the alkene. This transition state agrees with a model proposed by Houk.

A final example of the use of carbohydrate-derived dienophiles in Diels-Alder reactions involves the role of the carbohydrate as a chiral auxiliary. Shing and co-workers have described the Diels-Alder reaction of arabinose-derived dienophile **31** with butadiene under Lewis acid catalyzed conditions (Scheme 5). A mixture of cycloadducts **32** with *R* and *S* configurations at the cyclohexane ring stereocenter was obtained in 68% yield. The ratio was 73:27 in favor of the *R*-product when a benzyl glycoside was used; lower selectivities were observed with the corresponding methyl glycoside (*17*).

Carbohydrate Dienes. One of the earliest reported cycloaddition reactions of a carbohydrate diene was the self-condensation of a methyl 3-deoxy-3-nitrohexopyranoside derivative to give a bridged bicyclic product . Heating acetylated nitro sugar **33** resulted in the formation of a diene by loss of acetic acid, and cycloaddition occurred to give **35**. These products were converted to compounds with the 7-nitroisochromene system, in which the carbocyclic ring is aromatic (*18*).

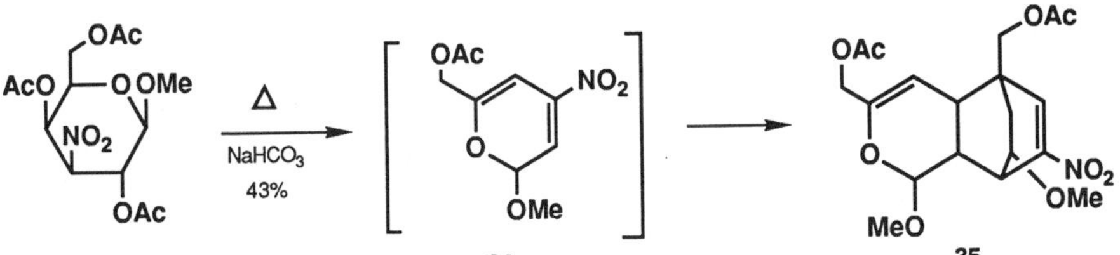

Scheme 4

Diels-Alder Reaction of a Quinone Methide with an Acyclic Sugar Dienophile

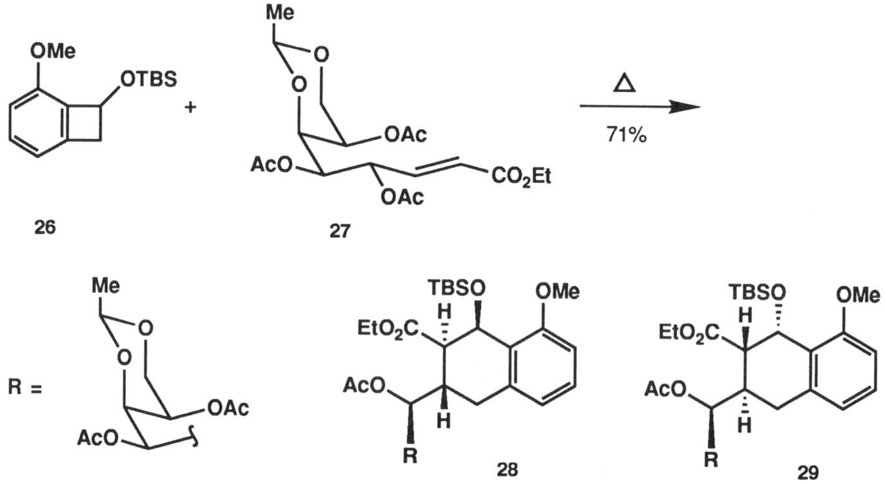

OMe
OTBS
+
Me
O
O
OAc
AcO
OAc
CO_2Et
Δ
71%
26
27

R =
Me
O
O
OAc
AcO

TBSO
OMe
H
EtO_2C
AcO
H
R
28

TBSO
OMe
H
EtO_2C
AcO
H
R
29

Scheme 5

Carbohydrate Chiral Auxiliaries

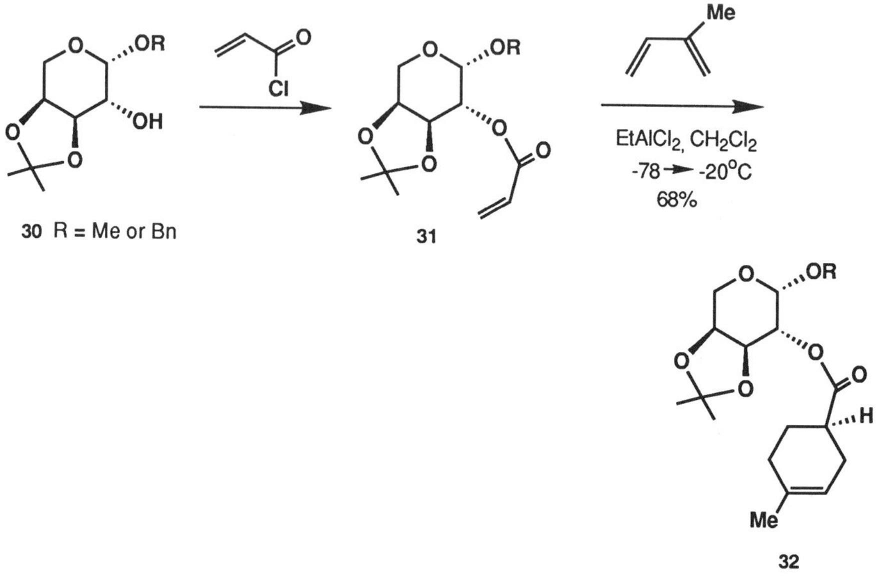

O
OR
OH
O
O
30 R = Me or Bn

O
Cl

O
OR
O
O
O
O
31

Me

$EtAlCl_2$, CH_2Cl_2
-78 → -20°C
68%

O
OR
O
O
O
O
H
Me
32

The synthesis of enantiomerically pure carbocycles from carbohydrate-derived dienes was first developed in the laboratories of Fraser-Reid, using dienes derived from 1,2:5,6-di-*O*-isopropylidene- α-**D**-glucofuranose, "diacetone glucose" (*19-22*). Three examples of Diels-Alder reactions of diacetone glucose-derived dienes are presented in Scheme 6. In entries (a) and (b), the cycloaddition of maleic anhydride anhydride occurred exclusively from the β-face of the diene in the endo mode, giving high yields of crystalline products. In (b) the anhydride ring in the cycloadduct undergoes ring opening to give an acid lactone which is isolated as its methyl ester. Multi-gram quantities of the ester were prepared and used to study further transformations. Entry (c) shows an intramolecular Diels-Alder reaction of an ester derived from the diene alcohol **38**. A mixture of cycloadducts **41** and **42** was obtained in which the major product resulted from migration of the double bond. The isopropylidene group in the dienes **36, 38**,and**40** directs the addition of the dienophile to the opposite face.

The synthesis and Diels-Alder reactions of dieno-pyranosides with general structure **43** was first described by Giuliano and Buzby (*23*). Annulated pyranosides **44** were obtained by the addition of maleic anhydride, maleimide, and *N*-phenylmaleimide (Scheme 7). Dieno-pyranoside **45** reacted with maleimide in refluxing benzene to give a single cycloadduct **46** in a recrystallized yield of 53% . Of the four possible products, the one formed results from the β-face addition of maleimide to **45** in the endo mode. In an effort to determine the effects of pyranose ring substitution on the stereoselectivity of cycloaddition, a series of six dieno-pyranosides were synthesized and their Diels-Alder reactions were investigated (*24*). The study revealed that, although the diastereofacial selectivity is influenced by the allylic substituent at C-3 on the pyranose ring, it is the anomeric substituent that exerts the larger directing effect in the Diels-Alder reactions. Dieno-pyranoside **47**, in which the allylic and anomeric groups occupy opposite faces of the diene, gave a mixture of cycloadducts in which the major component **48** resulted from addition to the β-face (*syn* to the allylic group). The structure of **48** was determined by NMR and X-ray crystallographic analysis, and the result was consistent with those obtained for the other dieno-pyranosides. This trend is surprising when one considers that the allylic group is attached to the diene terminus while the anomeric group is remote to any newly created stereocenter. The authors suggested that cycloaddition occurs to the face of the diene opposite the anomeric group because of unfavorable steric interactions that would exist in the transition state for *syn* addition.

The synthesis and Diels-Alder reactions of dieno-pyranosides **50** and **51** have been described by Lopez, Lameignere, and Lukacs (*25,26*). Both of these dienes reacted with maleic anhydride and dimethylacetylene dicarboxylate to give exclusively the

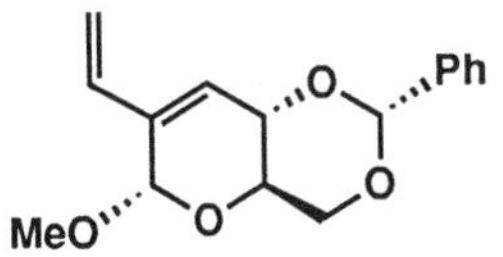

50 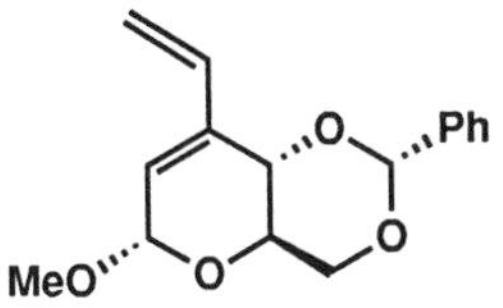 **51**

Scheme 6
Diels-Alder Reactions of Dienes Derived from Diacetone Glucose

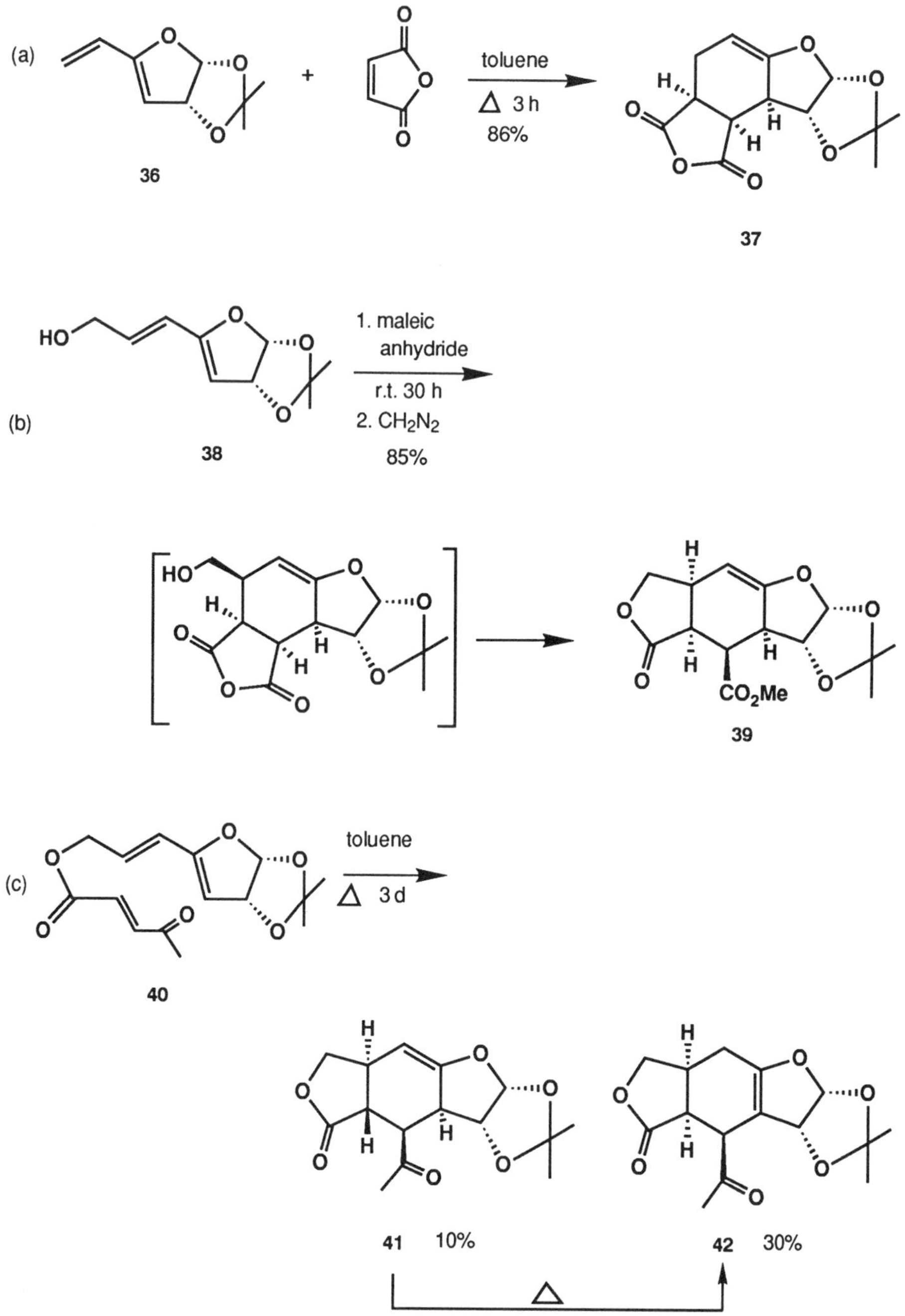

(a) 36 + maleic anhydride → toluene, Δ 3 h, 86% → 37

(b) 38 → 1. maleic anhydride, r.t. 30 h; 2. CH_2N_2, 85% → [intermediate] → 39 (CO_2Me)

(c) 40 → toluene, Δ 3 d → 41 10% + 42 30%

41 → Δ → 42

Scheme 7
Diels-Alder Reactions of Dieno-pyranosides

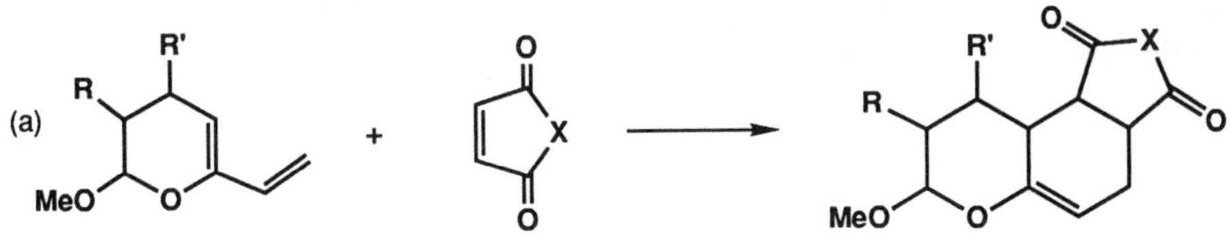

(a) 43 + → 44 X = O, NH, NPh

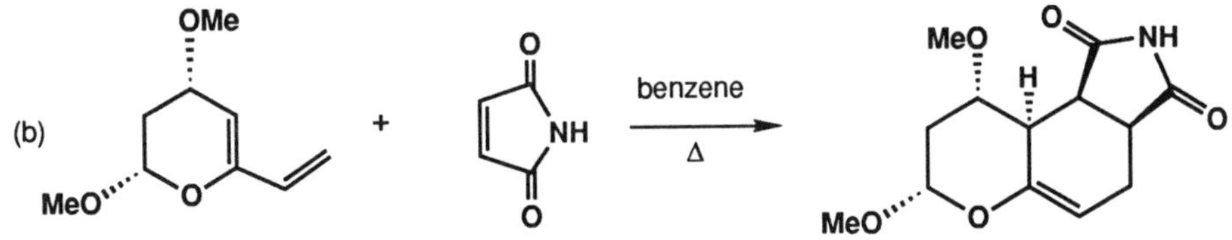

(b) 45 + → benzene Δ 46

(c) 47 NPM benzene Δ →

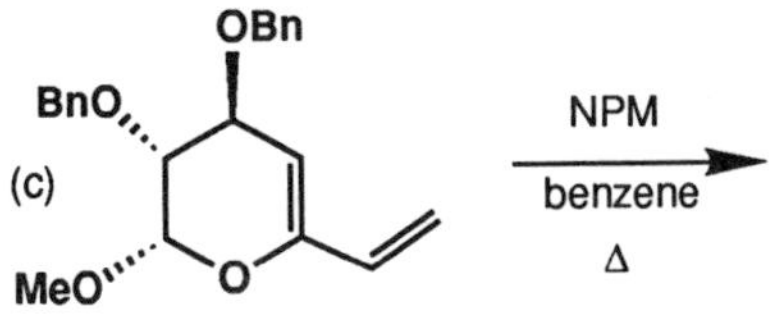

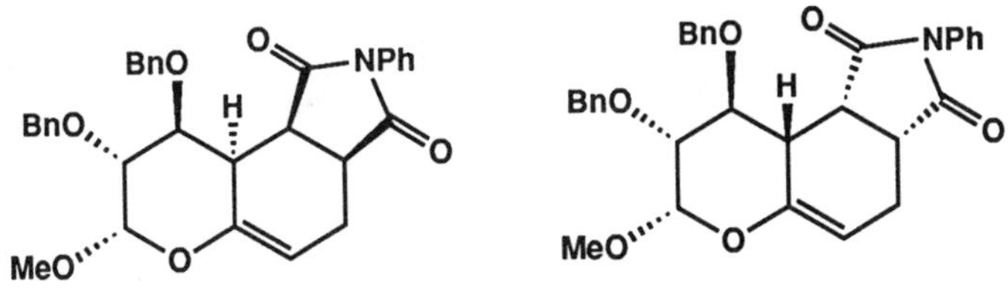

48 (β–endo) 6.4:1 49 (α–endo)

products resulting from addition of the dienophile to the face of the diene opposite the C-1 anomeric substituent, in the endo mode. Products are shown for **51**. Similar selectivities were observed by Lipshutz and coworkers in cycloadditions of **50** to both benzoquinone and naphthoquinone, the former giving a single cycloadduct and the latter a 10:1 mixture. These results were described in the context of a model study for the synthesis of anthraquinone antitmor agents (*27*).

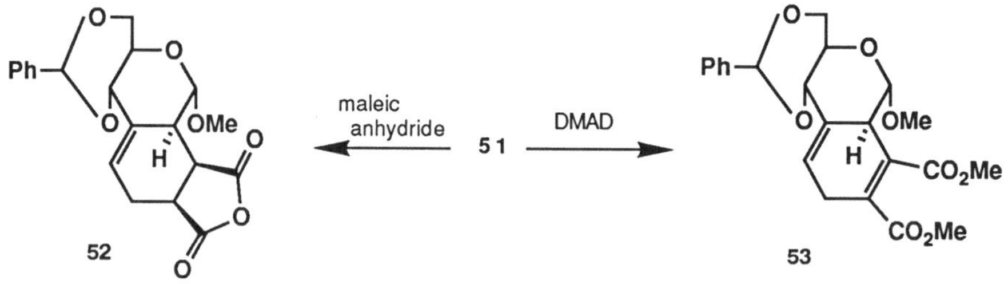

Acyclic carbohydrate-derived dienes have been described by Reitz, Jordan, and Maryanoff. Diene **54**, derived from arabinose, underwent cycloaddition to *N*-phenylmaleimide to give lactone **56** in a yield of 51% (*28*). A minor isomeric product was also obtained in 16% yield. The major product results from an intramolecular acylation of the initial endo cycloadduct. The stereoselectivity of the addition was found to be consistent with that reported in other studies involving alkoxy-substituted dienes.

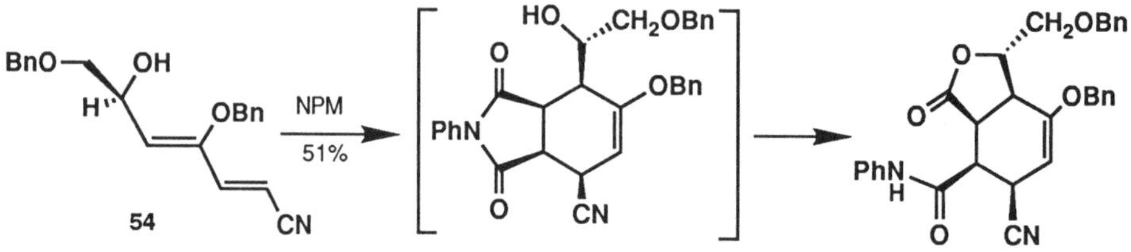

The use of carbohydrates as chiral auxiliaries in dienes has been described by Stoodley and co-workers (*29*). A series of glucopyranosyloxy-butadienes **57** with different diene substitution were synthesized and reacted with *N*-phenylmaleimide. Major cycloadducts resulted from addition of the dienophile to the diene in the endo mode (Scheme 8). A discussion of the stereoselectivity of this cycloaddition is included in the chapter by Franck in this volume.

Hetero Diels-Alder Reactions

Diels-Alder reactions in which carbon-heteroatom bonds are formed are useful for the synthesis of heterocyclic compounds. Many examplesof hetero Diels-Alder reactions are described in the recent monograph by Boger and Weinreb, including some which involve carbohydrates (*30*).

Gresham and Steadman reported the first example of a carbonyl group acting as a dienophile, and Kubler described the reaction of alkoxy butadienes with formaldehyde to give dihydopyran derivatives (*31,32*). Diene-aldehyde cycloaddition reactions were extended to carbohydrate substrates in the laboratories of David and co-workers. An example of their work in this area is the cycloaddition of diene **58** with n-butylglyoxylate to give disaccharides (*33*). The two main products of the reaction are shown; the one formed in higher yield results from endo addition to the "top" face of the diene while the minor product is formed by endo addition to the bottom face. Two other products are formed from exo addition to the two diene faces. This methodology was applied to the synthesis of the blood group A antigenic determinant (*34*).

58 → (butyl glyoxylate, benzene, 60°) → 60% (b-L) + 18% (b-D)

Diene-aldehyde cycloadditions in which the carbohydrate contains the formyl group were studied at high pressure by Jurczak and co-workers (*35*). Cycloaddition of 1-methoxybutadiene to aldehyde **59** at 20 kbar and 53° C occurred to give **60** with complete stereoselectivity. When the reaction was carried out at 11 kbar in the presence of $Eu(fod)_3$ as a catalyst, a 98:2 ratio of the cycloadducts was obtained in which **60** was the major product. The diastereoselectivity of the reaction is consistent with a Felkin-Anh transition state in which the diene approaches the formyl group from the less hindered face, in the endo mode.

Scheme 8
Glucopyranosyloxy-butadienes in Asymmetric Diels-Alder Reactions

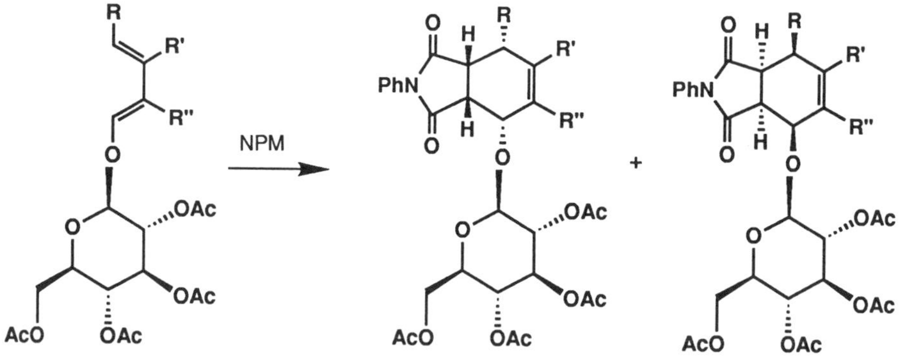

R R' R'' O OAc AcO OAc OAc 57 NPM O PhN H R R' R'' O OAc AcO OAc OAc major + minor

Scheme 9
Synthesis of Octosyl Acid A

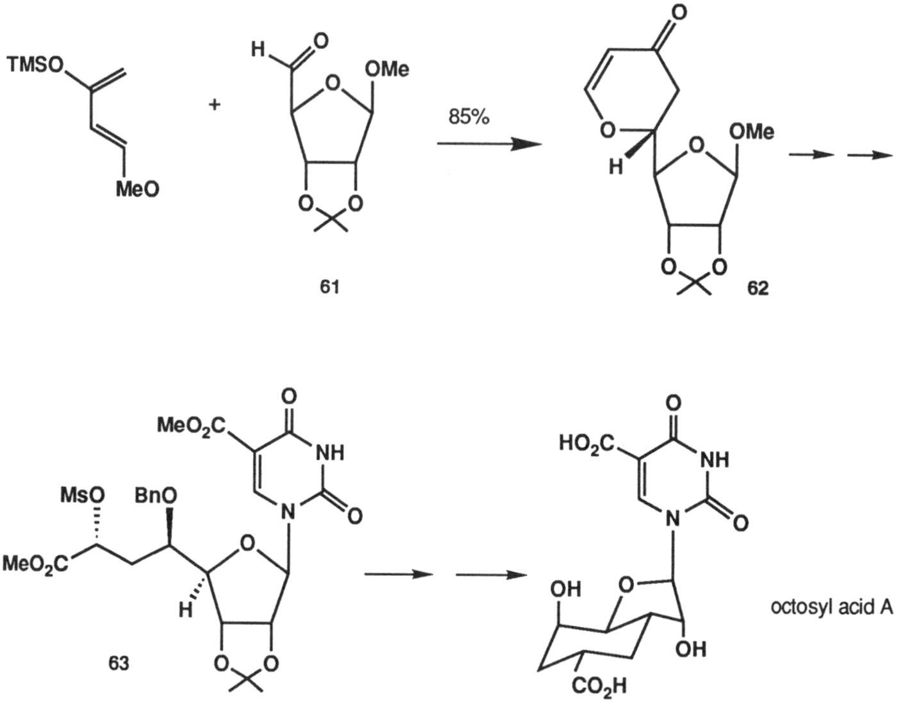

TMSO MeO + H O O OMe O O 61 85% O O OMe H 62

MsO BnO MeO_2C MeO_2C O NH N O H 63 O O HO_2C O NH N O OH O OH CO_2H octosyl acid A

59 + OMe → 20 kbar 53°C Et$_2$O 72% → 60

The diene-aldehyde cycloaddition has been further developed in the laboratories of Danishefsky and co-workers, where significant advances in methodology have been demonstrated by many successful applications to the synthesis of natural products (*36-38*). Two examples in which carbohydrate-derived substrates are utilized in the cycloaddition step are the synthesis of octosyl acid A and the synthesis of spectinomycin. In the synthesis of octosyl acid A (Scheme 9), Danishefsky's diene reacted with aldehyde **61** to give **62** in 85% yield (*39*). The stereoselectivity of the addition to **61** was that predicted by Cram's rule. For the synthesis of spectinomycin, cycloaddition of diene **64** to acetaldehyde occurred in the presence of Eu(fod)3 to give a 5.7:1 mixture of two products. The major diastereomer was the one desired for the spectinomycin synthesis (*40*).

64 → CH_3CHO, Eu(fod)$_3$ → 5.7:1

Although inverse-electron demand Diels-Alder reactions are uncommon in the carbohydrate literature, a recent example reported by Chapleur and Euvard illustrates that this type of reactivity is accessible to certain carbohydrate derivatives (*41*). The addition of a vinyl ether to **65** gave a 78% yield of cycloadduct **66**.

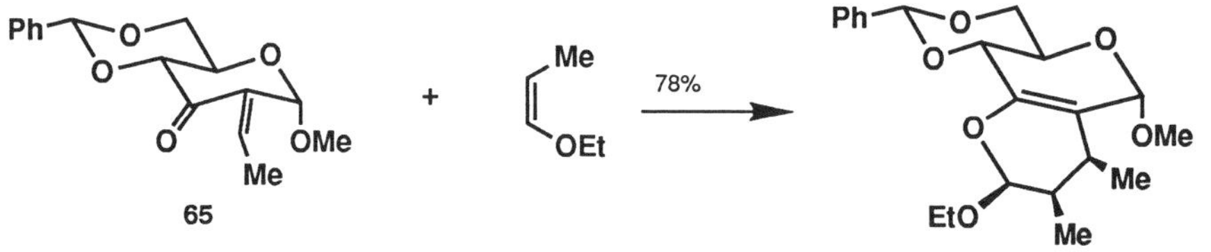

65 + → 78% → 66

Hetero Diels-Alder reactions involving chloronitroso sugars have been described by Vasella (*42*). A single cycloadduct was obtained in the reaction of **67** with cyclohexadiene. Compound **67** was obtained from **D**-mannose. The carbohydrate moiety was cleaved from the cycloadduct by *in situ* methanolysis to give **68**.

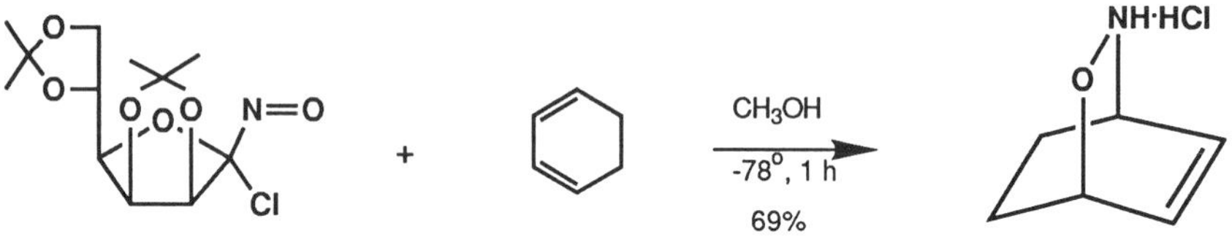

67 + cyclohexadiene → CH_3OH, -78°, 1 h, 69% → 68

Dipolar Cycloadditions

Dipolar cycloadditions are particularly useful for the synthesis of nitrogen-containing organic compounds. The heterocycle formed initially by dipolar cycloaddition may itself be a target, or the cycloadduct may be transformed by ring-opening to provide desired functionality in a cyclic or acyclic target. In carbohydrate chemistry, dipolar cycloaddition reactions have been used mainly for the synthesis of amino sugars such as daunosamine from non-carbohydrate precursors; however, reactions involving carbohydrates as 1,3-dipoles or dipolarophiles have also been described. Carbohydrate-derived nitrones have been studied extensively in the laboratories of Vasella and co-workers, primarily in cases where the carbohydrate functions as a chiral ausiliary. Representative examples of their work in this area are presented below.

Early studies of carbohydrate nitrone cycloadditions afforded isoxazolidine nucleosides (*43*). For example, dipolar addition of ethylene to *N*-glycosyl nitrone **69**, formed in situ from t-butyl glyoxylate and the carbohydrate oxime (acyclic), gave cycloadduct **70** in 78% yield and 72% diastereoselectivity. Hydrolysis of **70** in acid resulted in cleavage of the carbohydrate moiety to give isoxazoline **71**, a proline analog. These studies were extended to include other dipolarophiles such as furan (*44*). Dipolar addition of nitrone **72** to furan gave **73** (Scheme 10). Osmylation of **73** followed by protection of the resulting diol gave **74** in 40% overall yield. A multi-step sequence gave amino sugar **75**,which had been used in a previous synthesis of nojirimycin **76**.

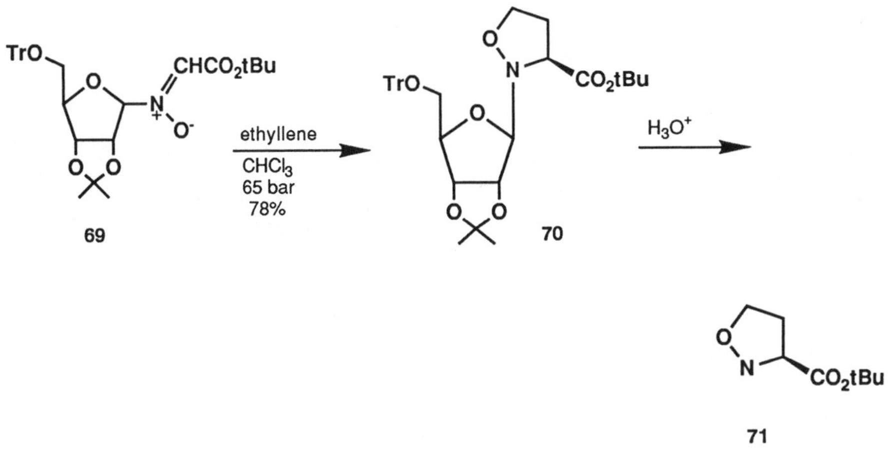

In more recent studies by Vasella, a rationale for the high diastereoselectivity of additions to *N*-glycosylnitrones has been described (*45*). The selective formation of the 5*S* isomer of **78** from **77** is thought to be the result of a kinetic anomeric effect. The reactions of the carbocyclic analogs of **77** proceeded in much lower stereoselectivity.

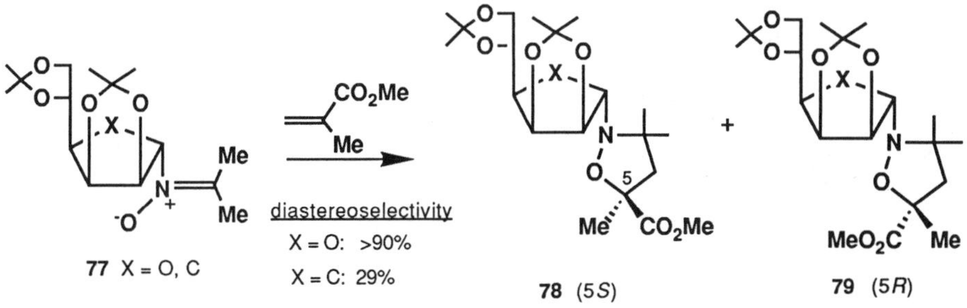

DeMicheli and co-workers have studied the facial selectivity of nitrile oxide cycloadditions to unsaturated sugars of type **80** (Scheme 11). In a theoretical treatment, minimum energy conformations of the alkene ground state were calculated using MM2, and these conformations were then used to evaluate energies of model transition states (*46*). The transition state model proposed by Houk, in which the "large" group on the allylic carbon is *anti* to the approaching nitrile oxide as shown, was found to be consistent with the product distribution.

Scheme 10

Synthesis of Nojirimycin

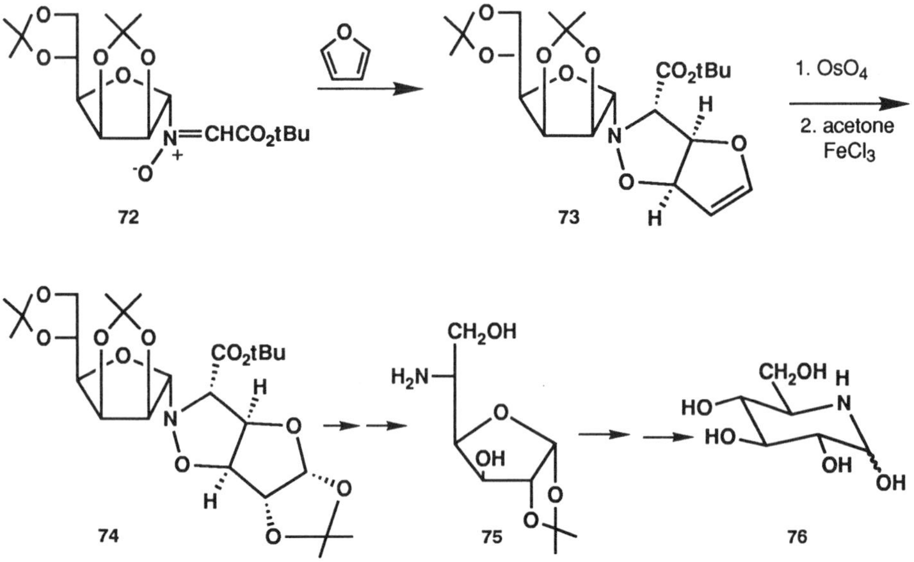

72 → 73 (furan); 1. OsO_4 2. acetone $FeCl_3$ → 74 → → 75 → → 76

Scheme 11

Nitrile Oxide Cycloadditions to Carbohydrate Alkenes

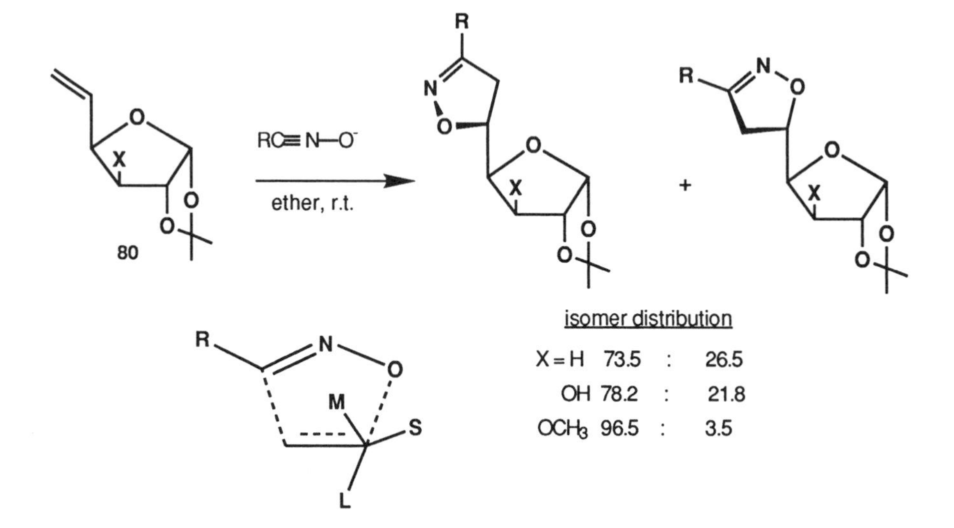

Regio- and steroselective dipolar cycloadditons of benzonitrile oxide and *C,N*-diphenylnitrone to levoglucosenone have been recently reported (*47*). In both cases, the major product results from the addition of the 1,3-dipole to the face of the enone opposite the 1,6-anhydro bridge, with the dipole oxygen becoming attached to the β-carbon of the enone. Facial selectivities of 100:1 were observed in the nitrile oxide cycloaddition).

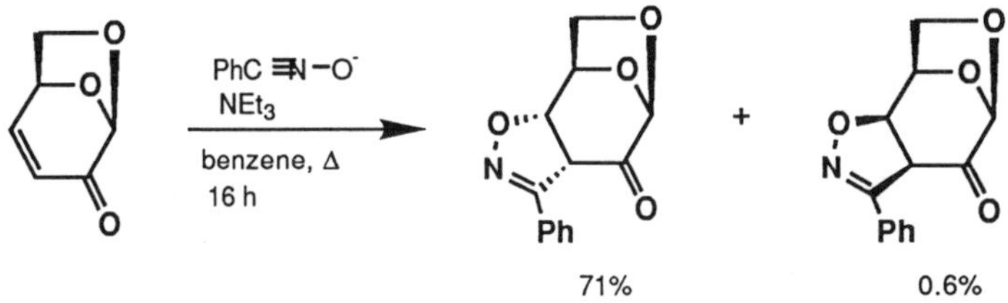
PhC ≡N−O⁻
NEt3
benzene, Δ
16 h
Ph
Ph
71%
0.6%

A synthesis of chiral hydroxypyrrolidines by Buchanan and co-workers is based on the intramolecular dipolar addition of an unsaturated sugar azide (Scheme 12). The *E* -ester **81** was converted to azide **82** via the 6-triflate. Upon heating, **82** underwent cycloaddition to give **83** in 68% yield. Treatment with base followed by catalytic reduction gave ester **85**. The 1-epimer of **85** was obtained by the same route when the *Z*-isomer of **81** was used (*48*).

Scheme 12

Synthesis of Chiral Pyrrolidines

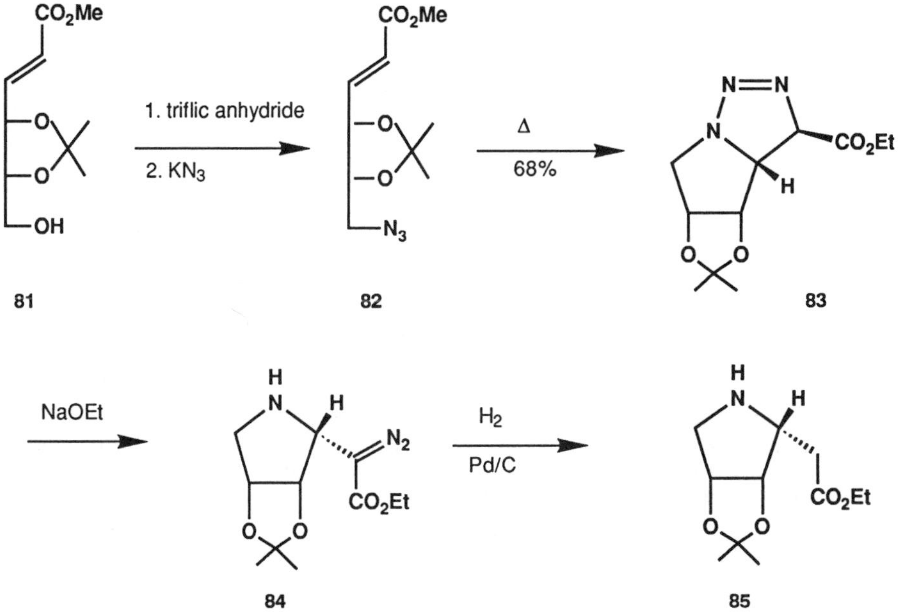
CO2Me
OH
81
1. triflic anhydride
2. KN3
CO2Me
N3
82
Δ
68%
CO2Et
83
NaOEt
N2
CO2Et
84
H2
Pd/C
CO2Et
85

Photochemical Cycloadditions

Photochemical cycloadditions have been among the least widely studied cycloaddition reactions of carbohydrate derivatives, however, some noteworthy examples have been described in the earlier as well as more recent lierature.. Photoannulations of pyranoid enones with both vinyl acetate and ethylene have been described by Fraser-Reid, Hicks, and Primeau (*49*). Cycloaddition of vinyl acetate to **86** gave the cyclobutano pyranoside **87** as a mixture of diastereomers. This reaction was explored as an approach to the asymmetric synthesis of the insect pheromone grandisol.

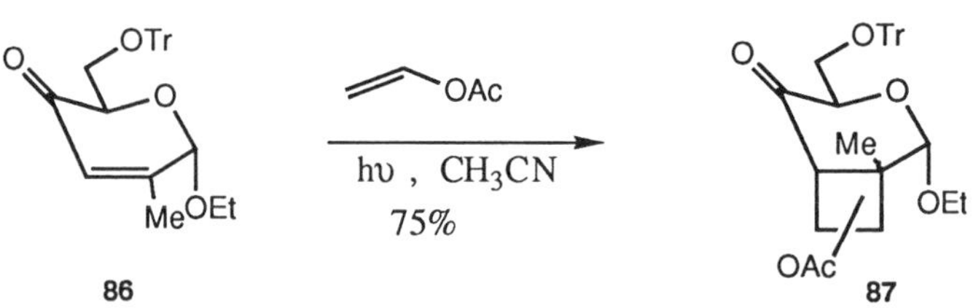

Photochemical reactions of 3,4,6-tri-*O*-acetyl-**D**-glucal and its 2-acetoxy derivative have been investigated in mixtures of acetone and 2-propanol by Ishido and co-workers (*50,51*). Irradiation of a solution of the glucal in a mixture of 9:1 acetone-2-propanol resulted in the photocycloaddition of acetone to the glucal double bond, giving oxetane **88**. The 2-acetoxy glycal was found to be less reactive toward cycloaddition, and gave mainly 1-hydroxy-1-methylethyl) radical adducts.

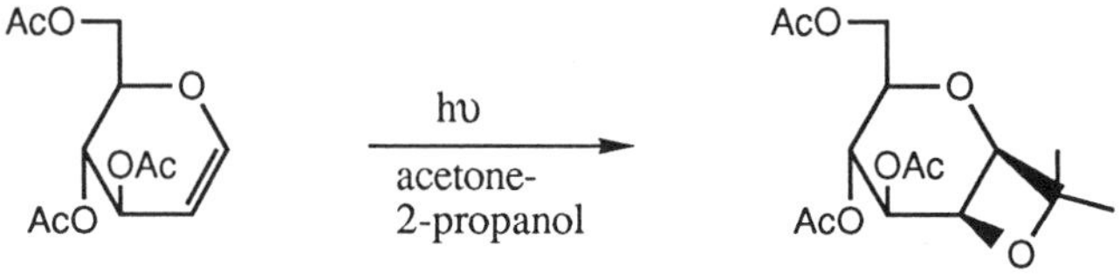

Carbohydrates have been used as chiral auxiliaries in asymmetric Paterno-Buchi reactions described by Scharf and co-workers (*52*). The diastereomeric excess obtained in the oxetane formation was found to depend heavily on both the reaction temperature and the structure of the carbohydrate auxiliary.

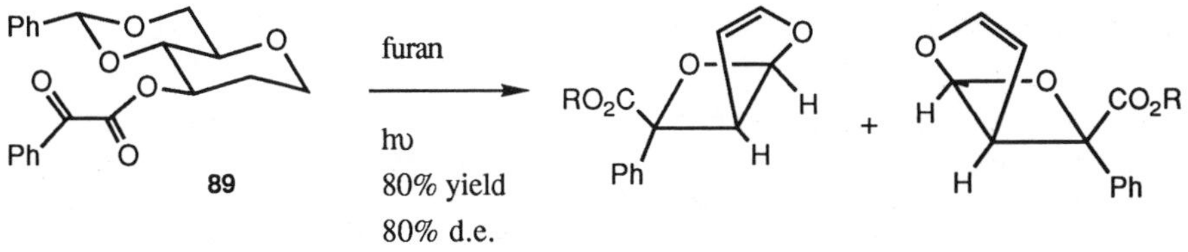

Ph, O, O, O, O, O, Ph, O, **89**; furan, hυ, 80% yield, 80% d.e.; RO_2C, O, O, H, H, Ph, +, O, O, H, CO_2R, H, Ph

In a recent study by Diaz and co-workers, the photoaddition products of glycosyl furans with singlet oxygen were transformed into nitrogen heterocycles *(53)*. Reduction of the singlet oxygen-glycosyl furan cycloadducts gave γ-diketones and γ-ketoesters which were transformed into substitted pyrazolines upon treatment with methyl diazoacetate (Scheme 13).

Scheme 13

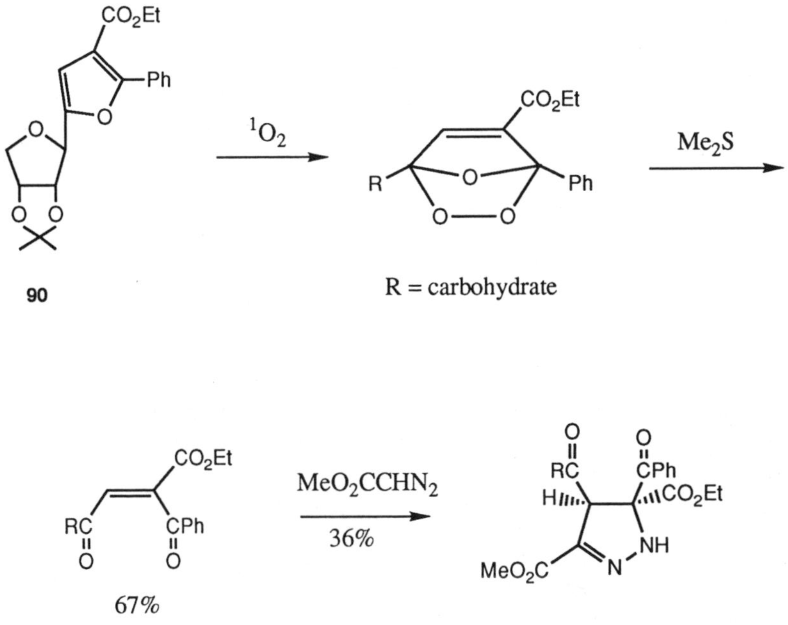

CO_2Et, Ph, O, O, O, O, **90**; $^1O_2$; CO_2Et, R, O, Ph, O, O, R = carbohydrate; Me_2S

CO_2Et, RC, O, CPh, O, 67%; MeO_2CCHN_2, 36%; O, RC, O, CPh, H, CO_2Et, MeO_2C, N, NH

Literature Cited

(1) Fraser-Reid, B.; Anderson, R.C. *Fortschr. Chem. Org. Naturst.* **1980**, *39*, 1.
(2) Fraser-Reid, B. *Acc. Chem. Res.* **1985**, *18*, 347; **1975**, *8*, 192.
(3) Holder, N.L. *Chem. Rev.* **1982**, *82*, 287.
(4) Rahman, M.A.; Fraser-Reid, B. *J. Am. Chem. Soc.* **1985**, *107*, 5576.
(5) Fraser-Reid, B.; Underwood, R.; Osterhout, M.; Grossman, J.A.; Liotta, D. *J. Org.Chem.* **1986**, *51*, 2152.
(6) Jurczak, J.; Tkacz, M. *Synthesis* **1979**, 42.
(7) Jurczak, J.; Bauer, T.; Kihlberg J. *J. Carbohydr. Chem.* **1985**, *32*, 2957.
(8) Chmielewski, M.; Jurczak, J. *J. Org. Chem.* **1981**, *46*, 2230.
(9) Card, P.J. *J. Org. Chem.* **1982**, *47*, 2169.
(10) Franck, R.W.; John, T.V. *J. Org. Chem.* **1983**, *48*, 3269.
(11) Franck, R.W.; Weinreb, S.N. in "Studies in Natural Products Chemistry, vol. 3, Stereoselective Synthesis (Part B)," Elsevier, Amsterdam, 1989, p. 173.
(12) Lichtenthaler, F.W. *Pure Appl. Chem.* **1978**, *50*, 1343.
(13) Dauben, W.G.; Kowalczyk, B.A.; Lichtenthaler, F.W. *J. Org. Chem.* **1990**, *55*, 2391.
(14) Bhate, P.; Horton, D. *Carbohydr. Res.* **1983**, *122*, 189.
(15) Horton, D.; Machinami, T.; Takagi, Y. *Carbohydr. Res.* **1983**, *121*, 135.
(16) Franck, R.W.; John. T.V.; Olejniczak, K.; Blount, J.F. *J. Am. Chem. Soc.* **1982**, *104*, 1106.
(17) Shing, T.K.M.; Lloyd-Williams, P. *J. Chem. Soc., Chem. Commun.* **1987**, 423.
(18) Baer, H.H.; Kienzle, F. *J. Org. Chem.* **1968**, *33*, 1823.
(19) Sun, K.M.; Fraser-Reid, B. *Synthesis* **1982**, 28.
(20) Sun, K.M.; Fraser-Reid, B. *J. Am. Chem. Soc.* **1982**, *104*, 367.
(21) Sun, K.M.; Giuliano, R.M.; Fraser-Reid, B. *J. Org. Chem.* **1985**, *50*, 4774.
(22) Fraser-Reid, B. Benko, Z.; Giuliano, R.M.; Taylor, N. *J. Chem. Soc., Chem. Commun.* **1984**, 1029.
(23) Giuliano, R.M.; Buzby, J.H. *Carbohydr. Res.***1986**, *158*, c1.
(24) Giuliano, R.M.; Buzby, J.H.; Marcopulos, N.; Springer, J. *J. Org. Chem.* **1990**, *55*, 3555.
(25) Lopez, J.C.; Lameignere, E.; Lukacs, G. *J. Chem.Soc., Chem. Commun.* **1988**, 706.
(26) Lopez, J.C.; Lameignere, E.; Lukacs, G. *J. Chem.Soc., Chem. Commun.* **1985**, 514.
(27) Lipshutz, B.H.; Nguyen, S.L.; Elworthy, T.R. *Tetrahedron* **1988**, *44*, 1988.
(28) Reitz, A.B.; Jordan, Jr., A.D.; Maryanoff, B.E. *J. Org. Chem.* **1987**, *52*, 4800.
(29) Larsen, D.S.; Stoodley, R.J. *J. Chem. Soc. Perkin Trans. I* **1989**, 1841.
(30) Boger, D.L.; Weinreb, S.M. "Hetero Diels-Alder Methodology in Organic Synthesis,"Academic Press, New York, 1987.
(31) Gresham, T.L.; Steadman, F.R. *J. Am. Chem. Soc.***1949**, *71*, 737.
(32) Kubler, D.G. *J. Org. Chem.* **1962**, *27*, 1435.
(33) David, S.; Eustache, J.; Lubineau, A. *J. Chem. Soc. Perkin Trans. I* **1974**, 2274.
(34) David, S.; Lubineau, A. Vatale, J.M. *New. J. Chem.* **1980**, *4*, 547.
(35) Jurczak, J.; Bauer, T.; Jarosz, S. *Tetrahedron* **1986**, *42*, 6486.
(36) Bednarski, M.; Danishefsky, S. *J . Am. Chem. Soc.* **1983**, *105*, 6968.

(37) Danishefsky, S.; Larson, E.; Askin, D.; Kato, D. *J. Am. Chem. Soc.* **1985**, *107*, 1246.
(38) Danishefsky, S.J.; DeNinno, S.L.; Chen, S.-h.; Boisver, L.; Barbachyn, M. *J. Am.Chem. Soc.* **1989**, *111*, 5810.
(39) Danishefsky, S.; Hungate, R. *J. Am. Chem. Soc.* **1986**, *108*, 2486.
(40) Danishefsky, S.; Aube, J.; Bednarski, M. *J. Am. Chem. Soc.* **1986**, *108*, 4145.
(41) Chapleur, Y.; Euvard,m M.-N. *J. Chem. Soc., Chem. Commun.* **1987**, 884.
(42) Felber, H.; Kresze, G.; Braun,H.; Vasella, A. *Tetrahedron. Lett.* **1984**, *25*, 5381.
(43) Vasella, A.; Voefray, R. *J. Chem. Soc., Chem. Commun.*, **1981**, 97.
(44) Vasella, A.; Voefray, R. *Helv. Chim. Acta.* **1982**, *65*, 1134.
(45) Huber, R.; Vasella, A. *Tetrahedron* **1990**, *46*, 33.
(46) DeAmici, M.; DeMicheli, C.; Ortisi, A.; Gatti, G.; Gandolfi, R.; Toma, L. *J. Org. Chem.* **1989**,*54*, 793.
(47) Blake, A.J.; Forsyth, A.C.; Paton, R.M. *J. Chem. Soc., Chem. Commun.* **1988**, 440.
(48) Buchanan, J.G.; Edgar, A.R.; Hewitt, B.D. *J. Chem. Soc. Perkin Trans. I* **1987**, 2371.
(49) Hicks, D.R.; Primeau, J.L.; Fraser-Reid, B. *Carbohydr. Res.* **1982**, *108*, 41.
(50) Araki, Y.; Senna, K.; Matsuura, K.; Ishido, Y. *Carbyhydr. Res.* **1978**, *64*, 109.
(51) Matsuura, K.; Araki, Y., Ishido, Y. *Bull. Chem. Soc. Jpn.* **1972**, *45*, 3496.
(52) Pelzer, R.; Jutten, P.; Scharf, H.-D. *Chem. Ber.* **1989**, *122*, 487.
(53) Diaz, R.R.; Calvo-Flores, F.G.; Guardia, L.A. *Carbohydr. Res.* **1989**, *191*, 209.

RECEIVED December 2, 1991

Chapter 2

Carbohydrate Dienophiles in [4 + 2] Cycloadditions

Richard W. Franck

Department of Chemistry and Institute for Biomolecular Structure and Function, Hunter College, City University of New York, 695 Park Avenue, New York, NY 10021–5024

The problem of diastereofacial preference of carbohydrate-derived dienophiles is discussed. Illustrative examples are chosen from our own research with three classes of dienophiles. The results are compared to other published data and are used to assess the predictive power of schemes grounded in MO theory. The most predictable face selectivity is observed in cyclic substrates where the facial differentiating elements are relatively fixed in the dienophilic carbon framework, e.g. glucals. Less predictability is found in acyclic sugar dienophiles where the precise geometry of the transition-state for the cycloaddition can only be determined by computation. The third class has the carbohydrate as a chiral auxiliary, attached to the dienophile through an ether link. Although useful diastereoselectivity is observed, the detailed understanding of the transition-state geometry for this last group of reactions has not been fully developed.

In the 60+ year history of the Diels-Alder reaction, the concept of diastereo- or enantiodifferentiation of the faces of dienes and dienophiles is fairly new.(*1*) My groups, first at Fordham University and then at Hunter College of the City University of New York have studied several variables in different diene and dienophile systems to try to understand what effects are important in stereodifferentiation in cycloadditions. This article will focus on our dienophile studies where carbohydrate derivatives serve as the diastereodifferentiators. Professor Horton discusses his work in the dienophile area in this volume and articles by Lubineau, Stoodley and Lopez will describe similar studies with dienes. Three different classes of dienophiles of interest to us will be described: acyclic structures exemplified by **1**, derived from sugars by Wittig chemistry, structures such as glucal **2** and lactone **3** where the dienophile double bond is within a ring, and vinyl glycosides **4** where the sugar is ether-linked to an achiral dienophile.

0097–6156/92/0494–0024$06.00/0

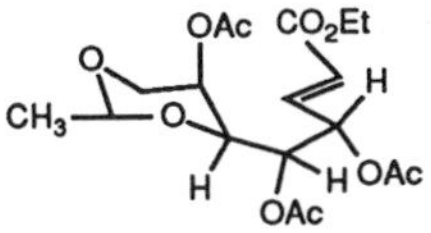

The conceptual basis for our studies in the acyclic series was a paper by Houk where he proposed a model for the cycloaddition of fulminic acid to propene based on MO calculations.(*2*) The model examined the interaction of the C-H bond of the allylic methyl of propene with the HOMO and LUMO of the transition-state and led to the enunciation of the following general principles : *attack of a reagent at an unsaturated site occurs such as to minimize antibonding secondary orbital interactions between the critical frontier molecular orbital of the reagent* and *those of the vicinal bonds and the tendency for staggering the vicinal bonds with respect to partially formed bonds is greater than for fully formed bonds*. With the appearance of that article, we extrapolated from the propene and fulminic acid of the calculations to the sugar-derived dienophiles shown in **Scheme 1** and their Diels-Alder reactions with ortho quinone methide **7** derived from the ring-opening of benzocyclobutene.(*3*) The major diastereomer in each addition was derived from attack on the *si* face of C-3 of the unsaturated sugar in which the configuration of the allylic carbon had been R (gluco) in the original pyranose starting material. This stereochemical relationship is denoted "unlike" according the Seebach-Prelog codification.(*4*).

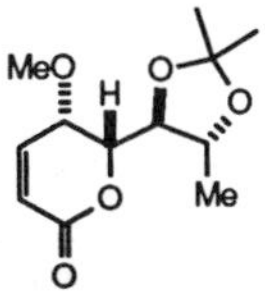

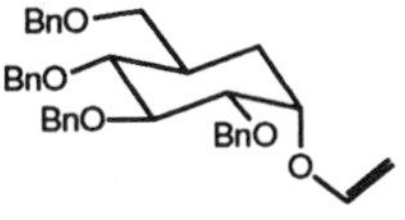

Scheme 1

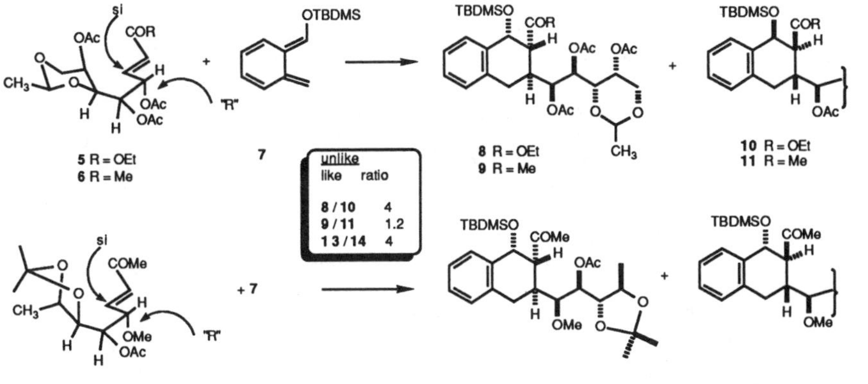

unlike/like	ratio
8 / 10	4
9 / 11	1.2
13 / 14	4

The unlike face selectivity was rationalized by simply considering the developing gauche interactions in the three like and three unlike staggered transition states shown in **Scheme 2.** Note that the activating and endo-directing group trans to the chiral function is omitted for clarity. Since the experiments afforded an unlike preference we focussed on models **A-C.** The model with the lowest repulsive gauche interactions was **B** with the alkoxy antiperiplanar to the developing bond. Next lowest was model **A** with an "inside"

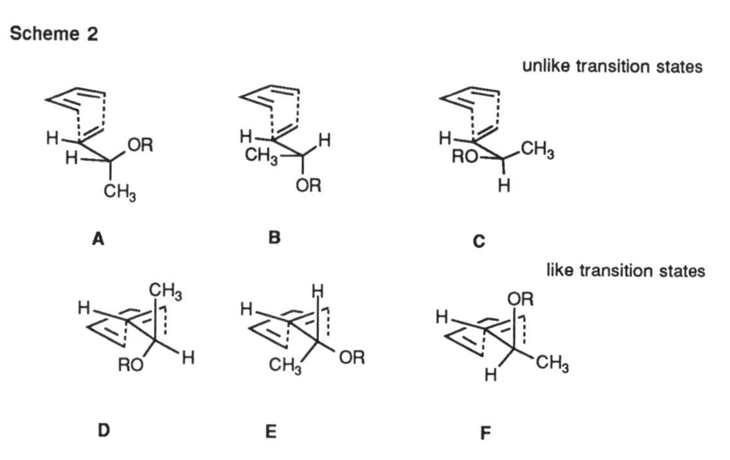

alkoxy group. After our initial work, Houk reported calculations which suggested that the inside alkoxy transition-state was favored in nitrile-oxide cycloadditions.(*5*) In the like series, it can be seen that model **D** has about the same value for gauche repulsions as the best unlike case **B** and the inside alkoxy case **E** is about the same as the inside unlike case **A**. Thus, we we were led to the conclusion that the antiperiplanar alkoxy group helped stabilize the transition-state by lowering the LUMO energy of the ground-state dienophile. Since the ortho quinone methide diene in our experiment was extremely reactive, an early reactant-like transition-state where ground-state effects still had influence seemed reasonable. With a confidence based on a limited data base, I extrapolated this analysis to allylically substituted dienes where it was postulated that an antiperiplanar oxygen would stabilize the HOMO of the diene, thus increasing the activation energy required to reach the transition-state. Hence our early observation of like products with substituted dienes was rationalized by invoking reaction via a diene analog of **D** where the allylic C-O bond was orthogonal to the bonds forming in the transition-state .(*6*) However, as results from other groups, as illustrated in Table 1 began to accumulate, it became clear that each dienophile-diene pair had its own unique transition-state and that a singular prediction of diastereofacial selectivity based principally on ground-state considerations of only one partner was not useful and often misleading.(*7*, *8*, *9*, *10*), Our more complete analysis of these arguments, applied to chirally substituted diene systems, has been published and it is almost certainly true that the same lack of generality obtains for dienophile systems derived from carbohydrates.(*11* , *12*)

Since the originating impetus for our Diels-Alder studies was a natural product structure where the like stereochemistry of structure **14 (Scheme 1)** was required from a fucose-derived dienophile, we resorted to a cyclic derivative of **12**, namely the lactone **25 (Scheme 3**). Here , the allylic and homoallylic substituents are *cis* and above the plane, hence in any plausible chair-like conformation of **25**, there must be an axial group blocking diene approach from the upper face and cycloaddition takes place cleanly from below to afford product **26** with the desired configuration.(*13*) In thought-provoking studies with cyclic enones, Danishefsky and Fraser-Reid have observed interesting effects with chiral substituents. Thus, dienophile **27** reacts with butadiene to afford adduct **28** in complete analogy to our results; but dienophile **29** affords syn adduct **30**. The missing experiment in this series is one with dienophile**31**.(*14*) With carbohydrate-derived enone **32**, it appears that the anomeric methoxy group, being axial, directs Diels-Alder reaction to the face opposite; yet,

Table 1 Face-selectivity results from other groups

dienophile	diene	conditions	like	unlike	R and X	ref.
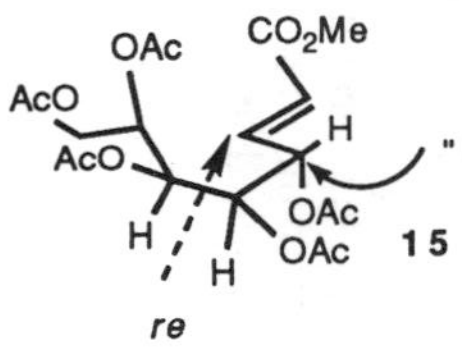 **15** (re)	 **16**	toluene reflux	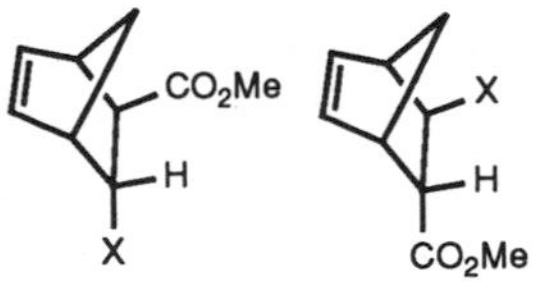 **17** / **18** 38%	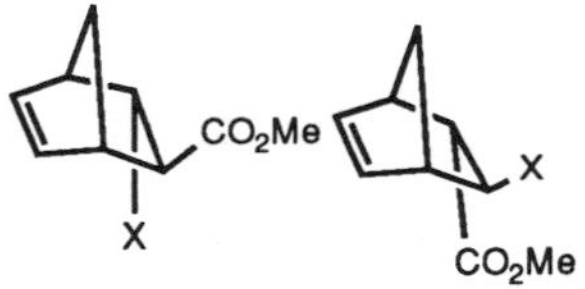 **19** / **20** 30%	X 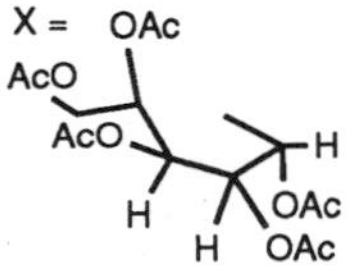 = (AcO, OAc, AcO, H, H, OAc, OAc, H)	7
15	**16**	$AlCl_3$, 0°	**17** / **18** 22%	**19** / **20** 57%	as entry above	7
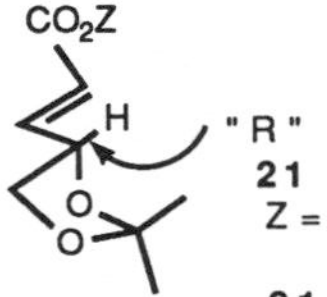 **21** Z = Me	**16**	benzene reflux	**17** / **18** 90%	---	X 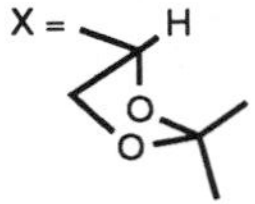 = (H, O, O)	8
21 Z = Me	**16**	Et_2AlCl CH_2Cl_2 -20°	**17** / **18** 93 %	---	as entry above	9
21 Z = t-Bu	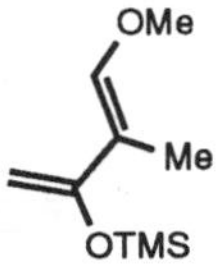 **22** (OMe, Me, OTMS)	115°	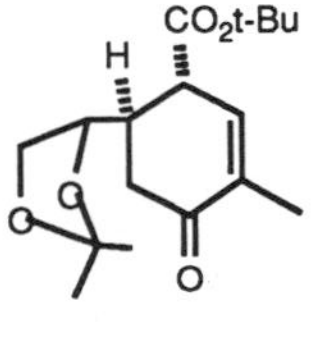 **23** 75%	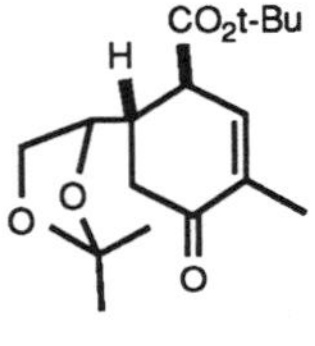 **24** 25%		10

Scheme 3

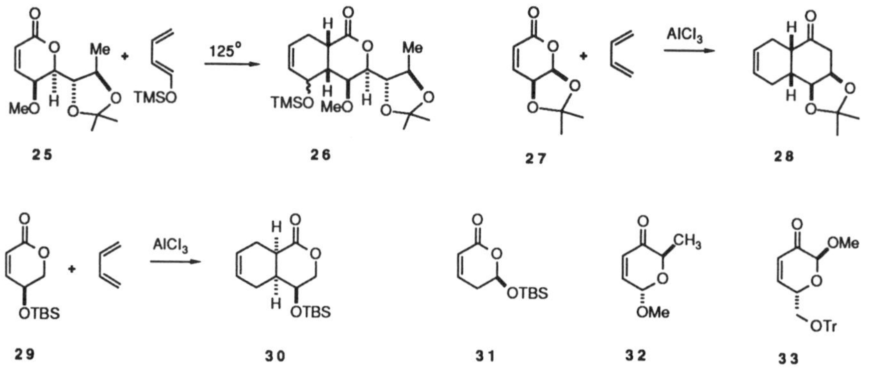

puzzlingly, cycloaddition with dienophile **33**, which appears simply to be a carbonyl-transposed version of **32,** fails completely.(*15*) Clearly, there is much yet to be learned about substituent effects on reactivity, let alone face-selectivity in carbohydrate-derived Diels-Alder systems.

It should be noted that the first instance of the use of a sugar-derived dienophile in a Diels-Alder reaction was reported by Baer in 1968.(*16*) Base treatment of nitroglucose derivative **34** produced the nitropyran **35** which functioned both as a diene and a dienophile to afford a major product the connectivity of which was tentatively assigned as shown in **36**, but the relative configuration of the product has not been assigned and the structure shown is simply a guess on the part of the present author. Interestingly, the product is racemic, presumably because the intermediate pyran **35** becomes racemic prior to cycloaddition.

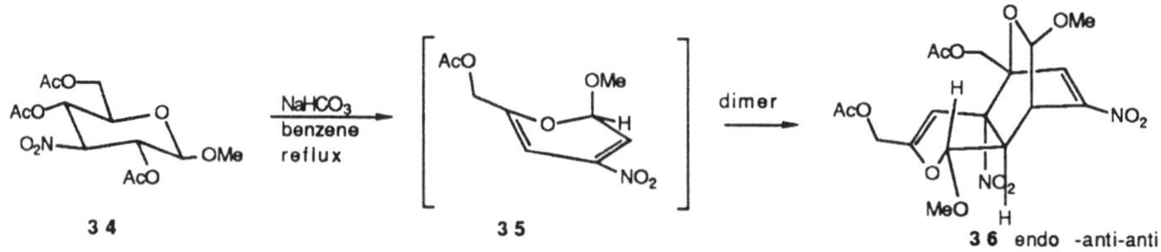

In an extension of cycloaddition studies designed to incorporate the chiral carbons of carbohydrates into non-carbohydrate natural product frameworks, we examined Diels-Alder reactions with glycals as dienophiles. Whereas our first attempts using benzocyclobutene-ortho quinone methide dienes (*17*) led to fairly complex product mixtures, the inverse-electron-demand Bradsher cycloaddition of isoquinolinium salts, illustrated in **Scheme 4**, gave cycloadducts in high yield with essentially 100% stereoselectivity. The initial cycloadduct **40** was usually processed under acetal-forming conditions to afford a tricyclic species such as **41.** In some cases, the initial adduct upon sequential acid and base treatment yielded a naphthaldehyde **42** with a highly functionalized side chain. In the glycals we examined, the stereochemistry of the adducts always arose via attack of the glucal double bond from the face opposite to both the allylic substituent at C-3 and the terminal carbon attached to C-5. However, a

Scheme 4

37 38 a) H^+ / H_2O b) NaOH 39 MeOH 40 H^+ / MeOH 41 42

43 stereochemistry of product always arises from this t.s.

44 stereochemistry of product never arises from this t.s.

37 + 45 → 46

simple examination of models, shown schematically in **43** and **44** didn't suggest that the C-3 or C-5 substituents, always equatorial, can really exert a strong stereoface directing effect. In fact, in a counter intuitive way, it appears that the axial proton at C-4 has a greater repulsive, hence opposite-face- directing effect than the quasiaxial proton at C-3 and the axial proton at C-5 combined Therefore, the model dienophile **45** was subjected to the reaction conditions and very high stereoface selectivity was observed. It is conceivable that the selectivity is due to some subtle steric interaction between the 6-substituent and the isoquinoline. Further experimentation will be required to separate the steric effect from a possible electronic effect. Interestingly, the photocycloaddition of azodicarboxylates to glycals, reported by the Merck Frosst group, displays a stereoselectivity controlled by the allylic function at C-3. Presumably, the Bradsher transition-state and the azo-cycloaddition transition-state have quite different bonding and charge-distribution.(*18*) Our method of fusing glycals onto an aromatic framework was used to accomplish an efficient total synthesis of the enantiomer of cryptosporin **49** from isoquinoline **47** and L-fucal **48**. Our synthesis confirmed the assignment of the relative configuration of the pyrano ring substituents, but it required the revision of the orignal assignment of the absolute configuration.(*19*)

47 + 48 10 steps 49 (-)-Cryptosporin

A different facet of our studies of the Bradsher cycloaddition was the use of carbohydrate-derived chiral auxiliaries in order to obtain face selective cycloadditions with isoquinoline salts.Our results are illustrated in Scheme 5. It is interesting to note that the α-anomer in our series gives very useful face selectivity. However, the rather lengthy sequence required for the removal of the auxiliary reduces the practical value of our method.(*20*)

Scheme 5

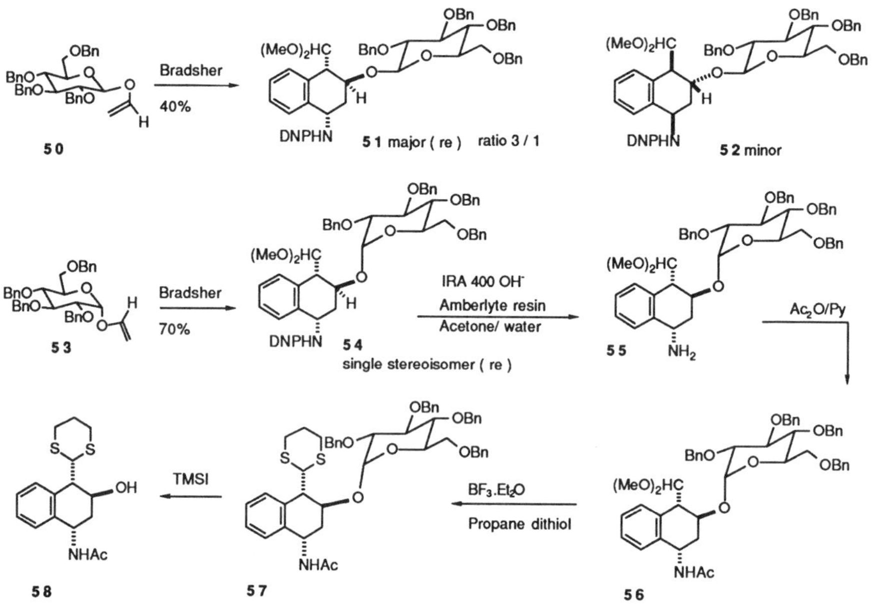

50 → Bradsher, 40% → 51 major (re) ratio 3 / 1 ; 52 minor

53 → Bradsher, 70% → 54 single stereoisomer (re) → IRA 400 OH^- Amberlyte resin, Acetone/ water → 55 → Ac_2O/Py → 56 → $BF_3.Et_2O$, Propane dithiol → 57 → TMSI → 58

Stoodley has described a rationalization for face selectivity in his extensive results with glycosyloxy butadienes.(*21*) The arguments, illustrated in **Scheme 6** are based on the assumption that the exo-anomeric effect which favors the ground-state conformation **59** also is operative in the Diels-Alder transition-state. Thus, in the β-glycosyl diene **60** the face exposed for attack has been observed to be the re face, consistent with the model. The model further predicts that in the α-series, diminished selectivity would be observed because neither face of the diene is clearly favored. As seen in our results, it is in fact the α-anomer which is the most selective, favoring attack on the re face. Thus as Stoodley points out in his most recent article ; " a combination of steric, electronic, and conformational factors are implicated in determining the diastereofacial selectivity of sugar-based dienes"and therefore one should not use a single model to generalize as to facial preference in the case of these rather flexible chiral auxiliary systems. However, the use of these systems is expanding rapidly, and as a body of results is accumulated, it may be possible to analyze the transition-states to determine what key structural factors are required for facial control. Recent examples which should be noted include the pioneering use by Posner of chiral auxiliaries for dienophiles in the inverse-electron-demand mode of cycloaddition, however not including carbohydrates as the chiral agent.(*22*) Stoodley (*21*) and David had used carbohydrates as chiral auxiliaries for dienes in Diels-Alder reactions of normal polarity.(*23*) Recently, Shing and Kunz reported on the use of carbohydrate auxiliaries for acrylate dienophiles in the normal Diels-Alder mode.(*24* , *25*)

Scheme 6. Conceptualization by Stoodley

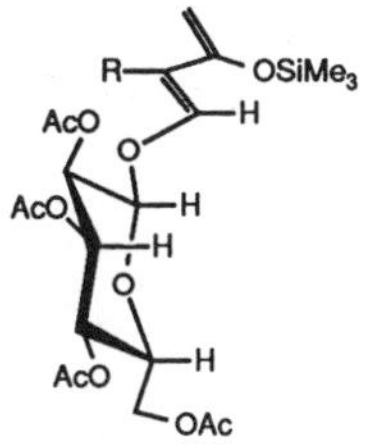

59
X-ray - ground state

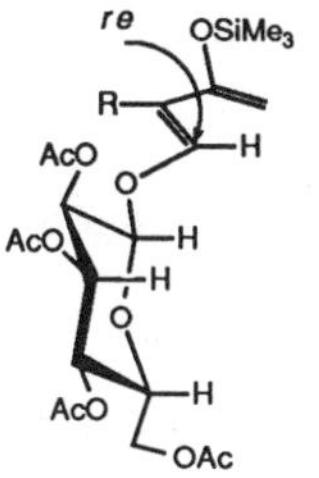

60
observed *re* (top) face attack if this is the reactive conformation in the T.S.

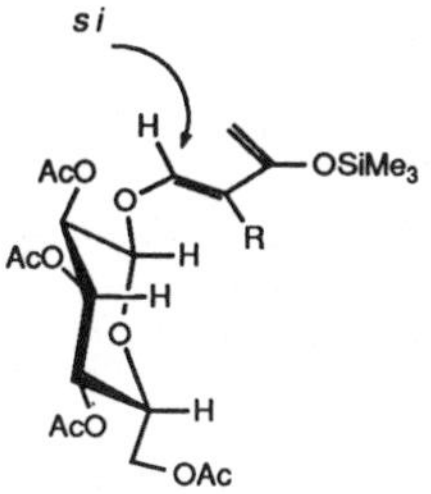

61
si (top) face attack disfavored since this rotamer is not populated because of R group interference with axial protons

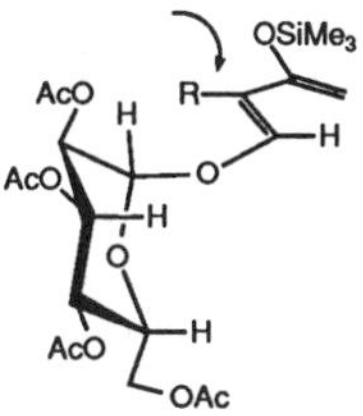

62

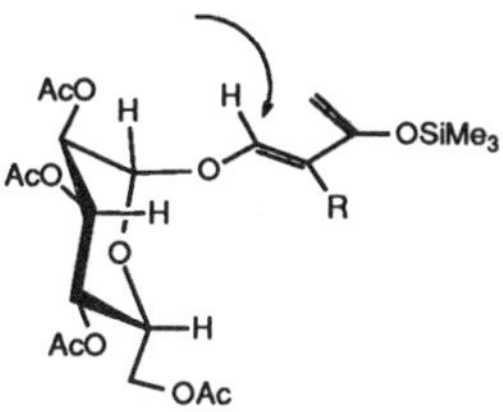

63

low face selectivity since neither rotamer is preferred

In conclusion, it can be seen that cycloadditions with carbohydrate-based dienophiles can lead to stereochemically complex carbocyclics in a straightforward way. The scope of the reaction is being expanded rapidly since the carbohydrate framework offers a variety of ways to construct dienophilic residues within the sugar carbon skeleton and also a range of linkages where non-carbohydrate dienophiles can be temporarily attached. The most challenging problem for the chemist is not the design of new dienophiles, but the development of a conceptual basis for prediciting the facial selectivity of the dienophile prior to the experiment.

Acknowledgement The experimental work in my group on face selectivity in cycloadditions, not all of which has been described in this article, was performed by A. Choudhury, S.C. Datta, R. Gupta, T. John, K. Olejniczak, C.S. Subramaniam and R. Tripathy. The X-ray structural analyses, necessary to prove configuration in many cases, were done by J. Blount, K. Onan, and G.J. Quigley. Theoretical calculations of transition-states have been carried out by J.J. Dannenberg and N. Kaila. I am indebted to all my collaborators for their input to this research. Funding for the work has come from the National Cancer Institute, the American Cancer Society, the Professional Staff Congress-CUNY grant program and the Division of Science and Mathematics of Hunter College / CUNY.

Literature Cited

1 Datta, S.C.; Franck, R.W.; Tripathy, R.; Quigley, G.J.; Huang, L:.; Chen, S-L.; Sihaed, A. *J. Am. Chem. Soc.* **1990,** *112*, 8472. Narasaka,K.; Iwasawa, N.; Inoue, M.; Yamada, T.; Nakashima, M.; Sugimori, J. *J. Am. Chem. Soc.* **1989,** *111*, 5340.
2 Caramella, P.; Rondan, N.G.; Paddon-Row, M.N.; Houk, K.N. J. Am. Chem. Soc. **1981,** *103,* 2438.
3 Franck, R.W.; John, T.V.; Olejniczak, K.; Blount, J.F. *J. Am. Chem. Soc.* **1982,** *104,* 1106.
4 Seebach, D.; Prelog, V. *Angew. Chem. Int. Ed. Eng.* **1982,** *21,* 654. It should be noted that the terms "anti" and "erythro" have been used to describe what we call "unlike" and that the strict application of the Seebach-Prelog nomenclature can create ambiguity. For example, in our cases the allylic chiral center was "R" in the original sugar, but has become "S" in the dienophile. Hence, descriptions in this field of research should have the *caveat* that only a comparison of 3-D models can insure the similarity or dissimilarity of the stereochemical sense of the reactions that are being compared.
5 Houk, K.N.; Duh, H-Y.; Wu, Y-D.; Moses, S.R. *J. Am. Chem. Soc.* **1986,** *108,* 2755.
6 Franck, R.W.; Argade, S.; Subramaniam, C.S.; Frechet, D.M. *Tetrahedron Lett.* **1984,** *25,* 3187.
7 Horton, D.; Machinami, T.; Takagi,Y. *Carbohydr. Res.* **1983,** *121,* 135.
8 Mulzer, J.; Kappert, M. *Tetrahedron Lett.* **1985,** *26,* 1631
9 Takano, S.; Kurotaki, A.; Ogasawara, K. *Synthesis* **1987,** 1075.
10 Lee, K-C.; Wu, J.C.C.; Yen, K-F.; Uang, B-J. *Tetrahedron Lett.* **1990,** *31,* 3563.
11 Kaila, N.; Franck, R.W.; Dannenberg, J.J. *J. Org. Chem.* **1989,** *54,* 4206.
12 Tripathy, R.; Franck, R.W.; Onan, K.D. *J. Am. Chem. Soc.* **1988,** *110,* 3257.
13 Franck, R.W.; Subramaniam, C.S.; John, T.V.; Blount, J.F. *Tetrahedron Lett.* **1984,** *25,* 2439.
14 Jeroncic, L.O.; Cabal, M-P.; Danishefsky, S.J.; Shulte, G.M. *J. Org. Chem.* **1991,** *56,* 387.
15 Fraser-Reid, B.; Underwood, R.; Osterhout, M; Grossman, J.A.; Liotta, D. *J. Org. Chem.* **1986,** *51,* 2152.
16 Baer, H.H.; Kienzle, F. *J. Org. Chem.* **1968,** *33* , 1823.
17 Franck, R.W.; John, T.V. *J. Org. Chem.* **1983,** *48,* 3269.
18 Leblanc, Y.; Fitzsimmons, B.J.; Springer, J.P.; Rokach, J. *J. Am. Chem. Soc.* **1989,** *111,* 2995.
19 Gupta, R.B.; Franck, R.W. *J. Am. Chem. Soc.* **1989,** *111,* 7668. An independent synthesis of cryptosporin, based on a nitroglycal annulation, with identical conclusions about the configuration of the natural product was reported shortly after our work: Briede, W.; Vasella, A.; *Helv. Chim. Acta* **1989,** *72,* 1649.
20 Choudhury, A.; Franck, R.W.; Gupta, R.B. *Tetrahedron Lett.* **1989** , *30,* 4921.
21 Gupta, R.C.; Larsen, D.S.; Stoodley, R.J.; Slawin, A.M.Z.; Williams, D.J. *J. Chem. Soc. Perkin Trans. 1* **1989,** 739. Beagley, B.; Larsen, D.L.; Pritchard, R.G.; Stoodley, R.J. *J. Chem. Soc. Perkin Trans. 1* **1990,** 3113.
22 Posner, G.H.; Wettlaufer, D.G. *J. Am. Chem. Soc.* **1986,** *108,* 7373.
23 David, S.; Lubineau,A.; Thieffry *Tetrahedron* **1978,** *34,* 299. David, S.; Eustache, J.; Lubineau, A. *J. Chem. Soc. Perkin Trans. 1* **1979,** 1795.
24 Shing, T.K.M.; Lloyd-Williams, P. *J. Chem. Soc. Chem. Commun.* **1987,** 423.

RECEIVED January 7, 1992

Chapter 3

Pyranose-Derived Dienes and Conjugated Enals

Preparation and Diels–Alder Cycloaddition Reactions

J. Cristóbal López[1] and Gabor Lukacs

Institut de Chimie des Substances Naturelles, Centre National de la Recherche Scientifique, 91198 Gif sur Yvette, France

Several pyranose derived dienes and pyranose derived conjugated enals were prepared and the stereochemical outcome of their cycloaddition reactions studied. The diene and the enal moiety were incorporated into the carbohydrate ring by different methods depending on the substitution pattern of the pyranose. The pyranosidic ring might act as a diene or dienophile in cycloaddition reactions, and as a 1-oxabutadiene system in hetero Diels-Alder reactions. The factors governing the stereoselectivities for each type of cycloaddition are rationalized. The role of the stereogenic allylic alkoxy substituents in the relative topicity of the reactions is analyzed.

The Diels-Alder reaction, more than 50 years after its discovery (*1*), is still one of the most efficient methods for the synthesis of carbocycles. A diene and a dienophile react to afford a carbocyclic compound with up to four stereogenic centers, in just one synthetic operation, and with a good control of regio- and stereochemistry (*2*). On the other hand carbohydrates, **I**, because of their nature have normally been transformed in polyoxygenated compounds (*3*). However, the last few years have witnessed an increasing interest in the not so straightforward transformation of carbohydrates into carbocycles (*4*).

Carbocycles from Carbohydrates Via Cycloaddition Reactions

Application of the Diels-Alder reaction to either carbohydrate derived dienophiles **II** or dienes **III** would constitute a good method to achieve the carbohydrate to carbocycle transformation, i.e., **I**→**IV**. The *synthetic efficiency* for this global transformation is going to be dependent on the *selectivity* in steps A and B (Figure 1). The efficiency in step A, has to do with the *regioselectivity* in the incorporation of the diene or dienophile unit into the carbohydrate from the starting pyranoside **I**; and in step B the *stereoselectivity* in the cycloaddition reaction has to be considered.

[1]Current address: Instituto de Química Orgánica General, C.S.I.C., Juan de la Cierva 3, 28006, Madrid, Spain

0097–6156/92/0494–0033$06.00/0

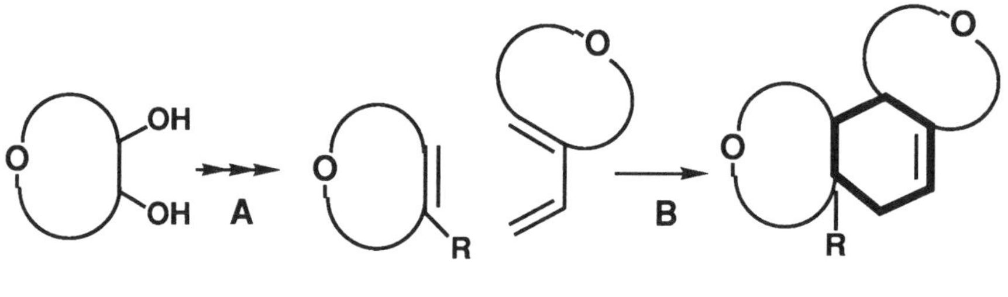

Figure 1

An additional advantage of sugar-based cycloadditions is that the carbocyclic products would be optically active. Although there is a loss of stereochemistry at two carbons in the pyranose ring in step A, are restored in the second step. The *overall result* of the transformation shown in Figure 1 is the stereocontrolled formation of two carbon-carbon bonds in the pyranose ring.

The Problem. We became interested in this subject when considering the feasibility of a carbohydrate based approach to biologically active compounds containing the bicyclofarnesol moiety. In our approach, (Figure 2) ring B would be formed through the Diels-Alder reaction of an appropriate pyranose substrate.

Retrosynthesis. The seminal work describing entry to such system had already been published by Fraser-Reid and Anderson (*3a*) (Figure 2). In their approach an α or β methyl substituted conjugated pyranosidic-ulose was the sugar precursor for the key intermediate "annulated pyranoside" (*5*). The carbohydrate behaves as a *dienophile*, and the origin of the quaternary methyl group in the bicyclofarnesol moiety (Figure 2) is the methyl substituent on the enone system.

We considered two alternative approaches to such systems, shown in Figure 2. In the *first approach* the pyranose ring would function as a *diene*, and the origin of the methyl on the quaternary carbon would be the methyl substituent in the dienic system.

In the *second approach* an homologated conjugated-enal could behave as the *dienophile* and the retrosynthetic origin for the methyl substituent in the target structure would be the formyl group activating the dienophile. Further interest in these approaches came from three points: first, the fact that no general synthetic routes to either of these systems could be found in the literature; second, that no cycloaddition reaction involving homologated conjugated pyranosidic enals was known; and finally that the carbohydrate unit had been only rarely used as a diene in cycloaddition reactions (*vide infra*).

Regioselective Access to α-Enal and Dieno-Pyranosides

In this context the initial task was to develop efficient routes from simple carbohydrates to dieno-pyranosides and pyranosidic α-enals (Figure 2).

Access to Dieno-Pyranosides. We were interested in developing synthetic schemes to dieno-pyranosides **V** (Figure 3a), and dieno-pyranosides **VI**. Systems type **V** are less reactive than their demethylated analogs **VI** (*6*) (the steric interaction of the methyl group with the vinyl residue causes the *s-cis* rotamer required for the cycloaddition reaction to take place to be unfavored); and the π-facial selectivity should, in principle, be the same.

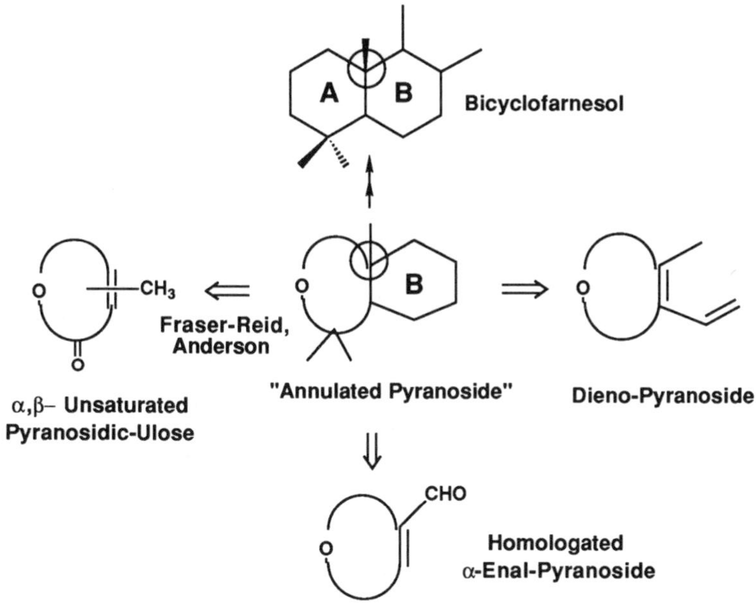

Figure 2

Vinyl Magnesium Route. A direct route to dieno-pyranosides (**V** or **VI**) is shown in Figure 3b, introduction of the vinyl residue is effected by reaction of vinyl magnesium bromide and a pyranoside-ulose **VII** (*7*), followed by elimination of the tertiary hydroxyl group in **VIII**. This route could, in principle, be common for both methyl, and desmethyl dienic systems.

Our target dienes **1**, **2**, and **3**, are shown in Figure 3c, and for the sake of simplicity we will be referring to them thorough out the text as "2,3-methyl-diene", "2,3-diene" and "3,2-diene". Both numbers in each name are used to define the location of the unsaturation in the pyranose ring, and the second of the two digits indicates the carbon atom in the pyranosidic ring to which the vinyl group is attached. Therefore compound **3** (Figure 3c) would be named "3,2 diene" because the unsaturation is located between positions 2 and 3 (carbohydrate numbering) and because the C-vinyl substituent is at C-2.

Access to the vinyl alcohol **4** (Figure 4a) is straightforward, and when elimination of the tertiary hydroxyl group was attempted by treatment with thionyl chloride in pyridine at -40°C, a mixture of three compounds was obtained. Along with the expected product **1**, an isomeric diene, **5**, was found in a 1:1 ratio. This compound arose from the elimination of the H-4 proton. The third component of this mixture was proved to be the rearranged allylic chloride **6** (*8*) formed *via* an S_NI' reaction of the intermediate allylic chlorosulphite (*9*).

Reasoning that torsional factors in the trioxa-trans-decaline moiety could have had an effect on the regioselectivity of the hydroxyl elimination (*10*), we prepared compound **7** from **4**. The benzylidene ring in **4**, was hydrolyzed in acetic acid solution, and the resulting free hydroxyl groups benzoylated to afford monocyclic product **7**. When compound **7** was subjected to the same thionyl chloride treatment only the product with the desired regiochemistry, **8**, was observed.

In an attempt to extend this *vinyl magnesium route* to the corresponding desmethyl systems (type **V**) compounds **9** and **10** were prepared (*11*). Unfortunately treatment of either **9** or **10**, with thionyl chloride led to complex reaction mixtures (Figure 4b). The same holds true for several isomers, possible precursors of 2,3 or 3,2-dienes (Figure 2c) (*11*).

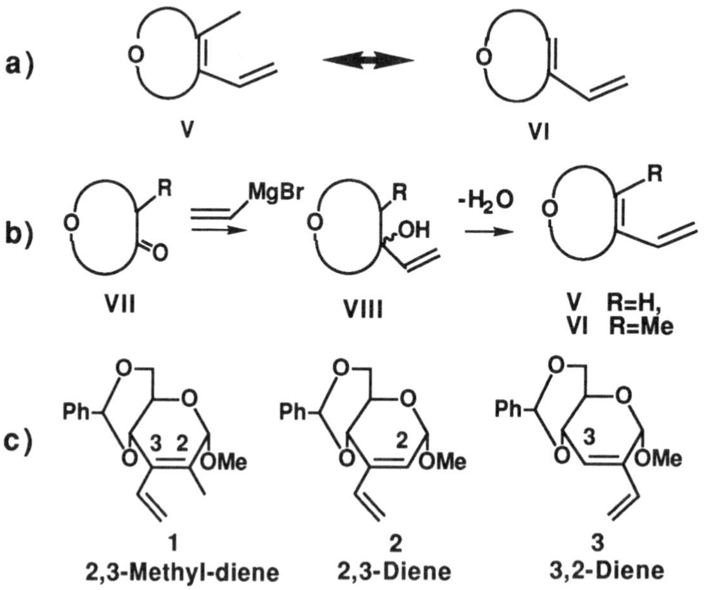

a) V ↔ VI

b) VII → (MgBr) VIII → ($-H_2O$) V R=H, VI R=Me

c) 1 2,3-Methyl-diene; 2 2,3-Diene; 3 3,2-Diene

Figure 3

Common Access to Pyranosido-Dienes and Conjugated α-Enals. It became obvious that the "Vinyl Magnesium Route" did not present any generality, and we were unable to find any rationalization for the regioselectivity of the elimination. In other words we could not discriminate between the two regioisomeric hydrogens H-a and H-b, as shown in Figure 5a, and furthermore, an allylic rearrangement process competed in

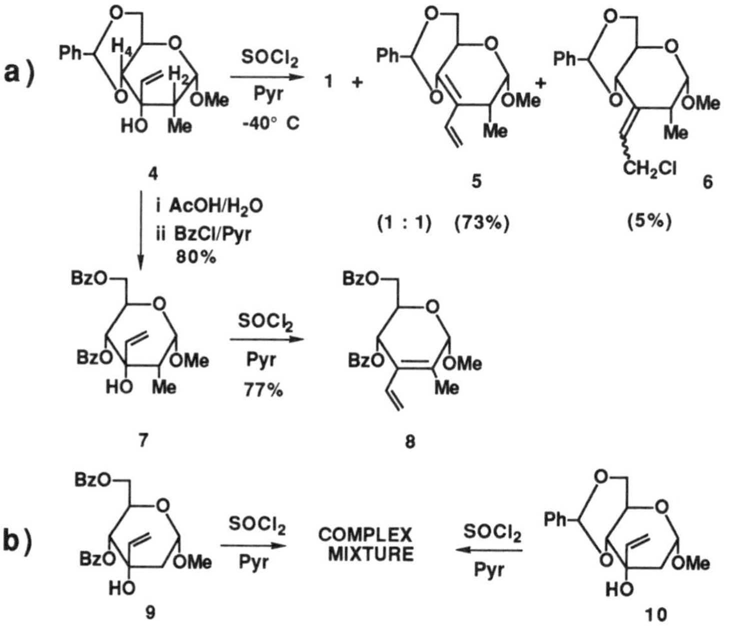

a) 4 ($SOCl_2$, Pyr, -40° C) → 1 + 5 (1 : 1) (73%) + 6 (5%)

4 → (i AcOH/H_2O, ii BzCl/Pyr, 80%) 7 → ($SOCl_2$, Pyr, 77%) 8

b) 9 → ($SOCl_2$, Pyr) COMPLEX MIXTURE ← ($SOCl_2$, Pyr) 10

Figure 4

several instances. On the other hand, we were interested in developing routes to both diene-pyranosides, and α-enal-pyranosides.

The need to devise a common protocol to diene-pyranosides and conjugated α-enals in which the regioselectivity of the elimination could be controlled, was the origin of the "dithiane route".

Dithiane Route. The route, outlined in Figure 5b fulfills these two requirements. An α-enal-pyranoside **IX** is correlated synthetically with a dieno-pyranoside **V**, and the former, **IX**, could be *regioselectively* obtained from a β-hydroxy-aldehyde, **X**, in which the hydroxyl group could be activated so that a regioselective β-elimination process could happen. Compound **X** could be correlated with a dithiane **XI** and further retrosynthesis led to an oxirane **XII** through 1,3 dithiane ring opening (*12*).

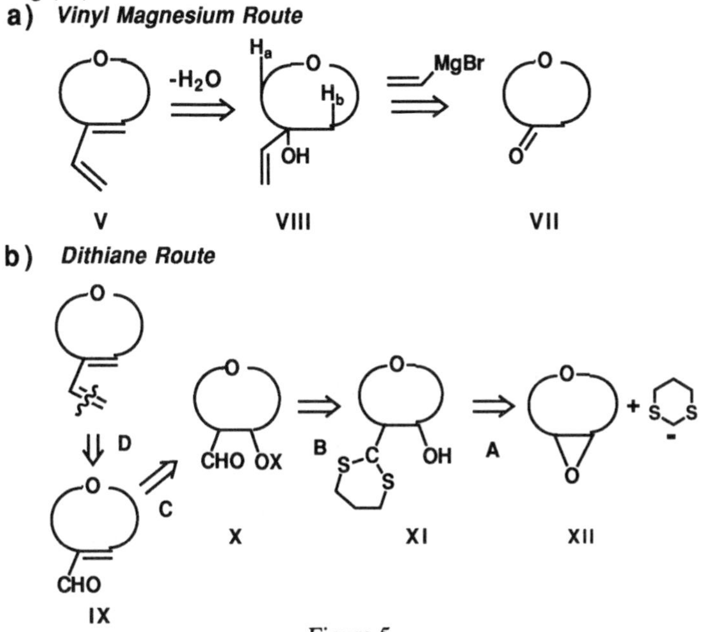

a) ***Vinyl Magnesium Route***

b) ***Dithiane Route***

Figure 5

In other words an epoxide, **XII**, could be the common precursor for both, dienes and conjugated enals by ring opening with lithium 1,3-dithianyl followed by a sequence that would include: deprotection, activation of the hydroxyl group, and β-elimination (Figure 5b).

We first explored this route in connection with the synthesis of the 3,2-enal **15** (Figure 6). Dithiane ring opening of the *allo*-epoxide **11** in tetrahydrofuran, without any catalyst, afforded one single isomer, **12**, (*13*) in 80% yield in gram scale operations. Unveiling of the dithiane group in compound **12** (*14*) did not afford any of the expected hydroxy aldehyde **13**, but a mixture of regioisomeric conjugated 1,2 and 3,2-enals, **14** and **15**, and in which 3,2-enal, **15**, was formed only as a minor isomer. A spontaneous elimination process was occurring opposite to the one desired. Nevertheless, and in keeping with our previous reasoning, enhancement of the leaving group properties of the substituent at C-3 reversed the observed regioselectivity of the elimination and led to **15**. When hydrolytic conditions (*14*) were applied to the acetyl derivative **16**, 3,2-enal **15** was obtained as the major isomer in 66% yield (Figure 6).

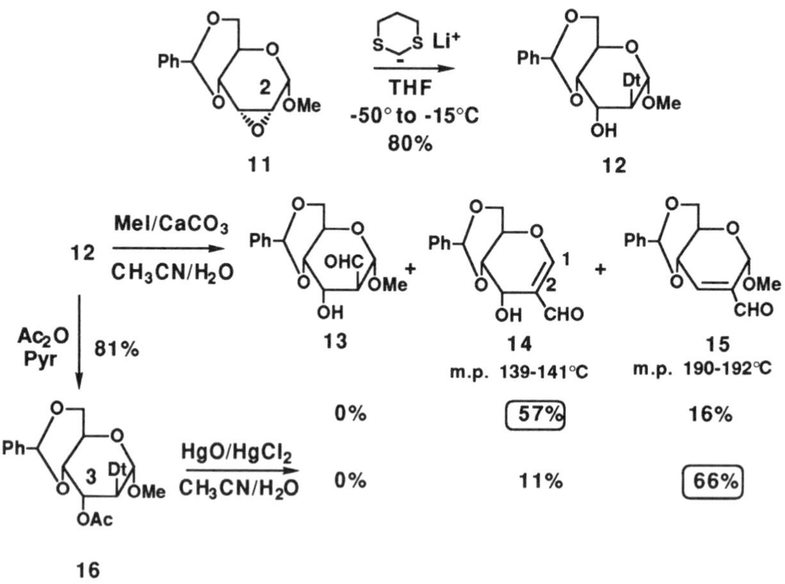

Figure 6

The same procedure allows access to 2,3-Enal **19** (*15*). Location of the formyl group at C-3 in enal **19**, required the use of the known *manno*-epoxide **17**. Treatment of **17** as above with dithianyl lithium, but this time in the presence of either HMPT or DMPU as catalysts, at higher temperature (-15°C -> +4°C), allowed us to obtain the regioisomeric dithianyl compound. After being activated as its mesylate, **18** on treatment with MeI/$CaCO_3$ in CH_3CN/H_2O, led directly to **19** as the sole product in 73% yield (Figure 7). Application of the "dithiane route" to the corresponding epoxides of the *galacto* series, **20** and **22**, led to the 2,3-; 1,2-; and 3,2-enals **21**, **23** and **24** respectively (*16*).

Preparation of Pyranosido-Dienes from α-Enal-Pyranosides. The retrosynthetic correlation between α-enals and diene-pyranosides shown in Figure 5b found precedent in the related work of Giuliano and Buzby (*17*). In fact α-enal pyranosides can be transformed into the corresponding dienes by Wittig alkenation using methyl triphenyl phosphonium bromide. As previously observed by Giuliano (*17*), Wittig reaction of these substrates proved to be quite sensitive to the base employed (Figure 8). Use of buthyl lithium to generate the ylide, gave only a 20% yield of the 1,2-diene **25** from the corresponding α-enal **23**, changing to potassium hexamethyldisilazide allowed the obtention of this diene in 70% yield. Dienes, **2**, **3** and **25**-**27**, shown in Figure 8, were prepared by reaction of the corresponding aldehyde precursors with methyl triphenyl phosphonium bromide in THF.

Enol-Ether Route. The dienes shown in Figure 8 contain a 4,6-di-*O*-benzylidene group, and that is related with an additional element of regiocontrol implied in the "dithiane route". We have shown how some regiocontrol could be exerted in the elimination step (C in Figure 5b) to afford one single regioisomer **IX** by enhancement of the leaving group properties of the 3-substituent in **X**. There is a second element for regiocontrol in this method, which is related to step A in Figure 5b. In fact regioselectivity in the ring opening of cyclic oxiranes such as **XII** is governed by the *Fürst-Plattner rule* for the trans-diaxial ring opening of oxiranes (*18*). The presence of

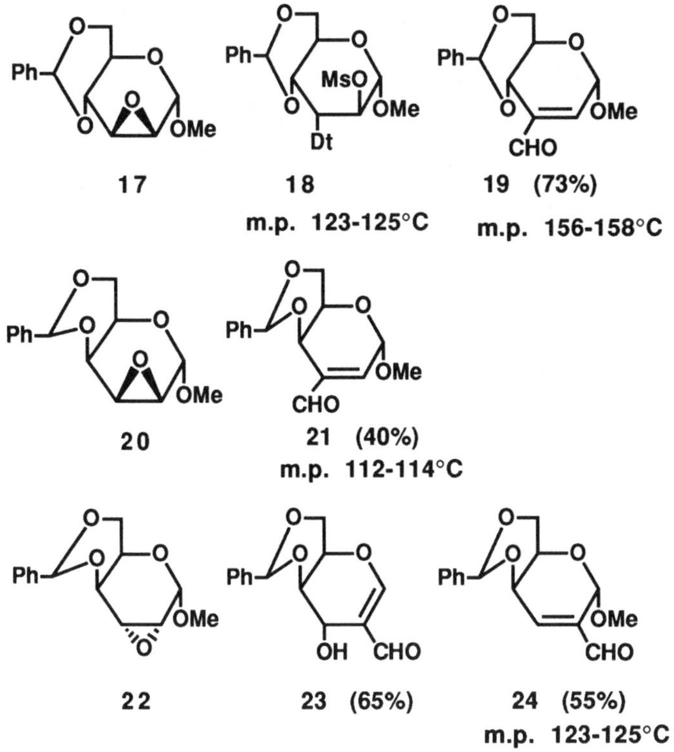

17

18
m.p. 123-125°C

19 (73%)
m.p. 156-158°C

20

21 (40%)
m.p. 112-114°C

22

23 (65%)

24 (55%)
m.p. 123-125°C

Figure 7

the benzylidene ring confers conformational rigidity to the pyranose moiety, and it is because of that conformational rigidity that only one regioisomer is obtained. Systems without the benzylidene ring have been shown to give mixtures of regioisomeric products *(19)* because, although the ring opening is always trans-diaxial, systems with conformational movility can react through different interconvertible conformers. For systems with conformational mobility we developed the so-called "enol-ether route"

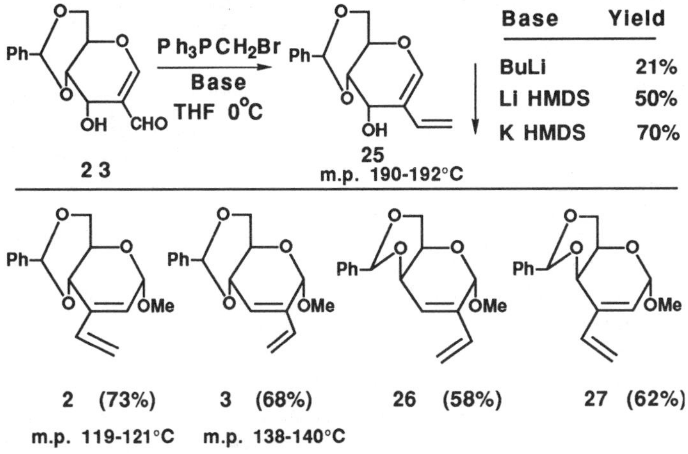

Ph_3PCH_2Br, Base, THF 0°C

23 → 25 (m.p. 190-192°C)

Base	Yield
BuLi	21%
Li HMDS	50%
K HMDS	70%

2 (73%) m.p. 119-121°C

3 (68%) m.p. 138-140°C

26 (58%)

27 (62%)

Figure 8

We found that conjugated-enals **IX** could also be obtained upon mild acidic treatment of α-alkoxy enol-ethers **XIII** (Figure 9a) derived from pyranoses *(20)*. The transformation can be carried out under two different experimental conditions: a) treatment of XIII with a 1M solution of pyridinium hydrochloride in pyridine (conditions A in Figure 9a), and b) treatment with a diluted acetic acid solution (B in Figure 9a). The yields in conjugated enals **28-31** are good, ranging from 70 to 90%, and the precursor enol ethers **XIII** can be efficiently obtained by Wittig-Horner reaction of the corresponding uloses *(21)*.

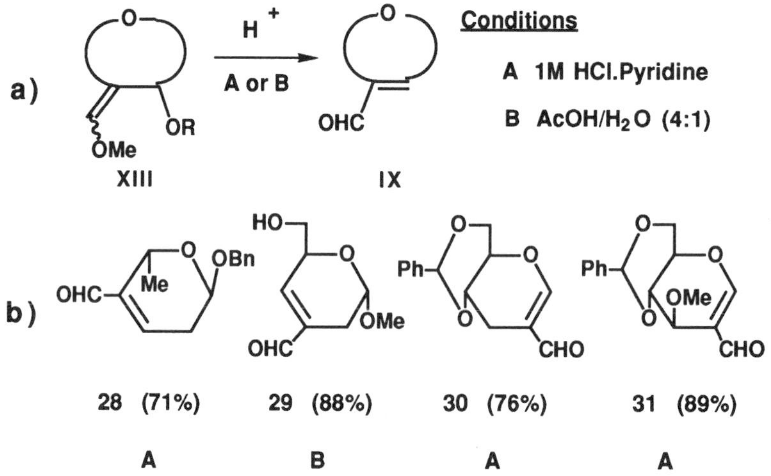

a) XIII → (H^+, A or B) → IX

Conditions

A 1M HCl.Pyridine

B $AcOH/H_2O$ (4:1)

b) 28 (71%) A; 29 (88%) B; 30 (76%) A; 31 (89%) A

Figure 9

Cycloaddition Reactions of α-Enals

The synthetic routes discussed above allow ready access to dienes and enals from simple carbohydrate precursors. In accordance with our initial plan we studied the stereochemical behavior of these systems in intermolecular cycloaddition reactions.

Homologated α-Enal Pyranosides as Dienophiles. 2,3-Enal pyranoside **19** reacted with an excess of butadiene in CH_2Cl_2 in the presence of $AlCl_3$ as catalyst at -78°C to give carbocyclic compound **32** (Figure 10a) *(8)* as a sole stereoisomer in 75-85% yield, arising from exclusive approach from the β face. Since all the carbohydrates described in this chapter belong to the D-series; α and β will be used thorough the text to denote the approach of the reagent, α face being the one opposite to the substituent at C-5). Unlike other dienophilic systems in carbohydrates *(22)*, $BF_3.Et_2O$ could also be used as a catalyst without affecting the yield.

On the other hand, reaction of the regioisomeric 3,2-enal-pyranoside, **15**, with butadiene took place under similar reaction conditions to afford also a single stereoisomer, **33**, in 75-85% yield, but this time the approach of the butadiene came from the opposite α-face. The stereochemistry for the stereoadducts was inferred from their ^{1}HNMR coupling constants, and by chemical correlation with a previously described Diels-Alder adduct *(23)* and has been discussed elsewhere *(8)*.

The cycloaddition reaction of α-enal pyranosides allows ready access to enantiomerically pure carbocyclic structures containing stereochemically defined quaternary carbon centers *(24)*.

Stereoselectivity. The direction of approach for the formation of **33** was explained in terms of an unfavorable interaction of the butadiene with the anomeric methoxy substituent in the transition state as shown in Figure 10b.

Our rationalization for the opposite behavior of enal **15** is shown in Figure 10b, and invokes a chelation complex between the formyl group at C-2 and the oxygen in the anomeric substituent which results in a conformational change of the pyranose ring so that the α-face would be easier to approach. Although such chelation, to give a six membered complex, would also be possible for **19** between the formyl group at C-3 and the oxygen at C-4, the conformational rigidity of the trioxa-trans-decaline system would preclude any conformational change.

a) Ph O O O OMe CHO 32 Ph O O O OHC OMe 33

b) Ph O O 4 O OHC 3 OMe 19 β-face attack Ph O O H H O 2 Me O Al O H α-face attack 15

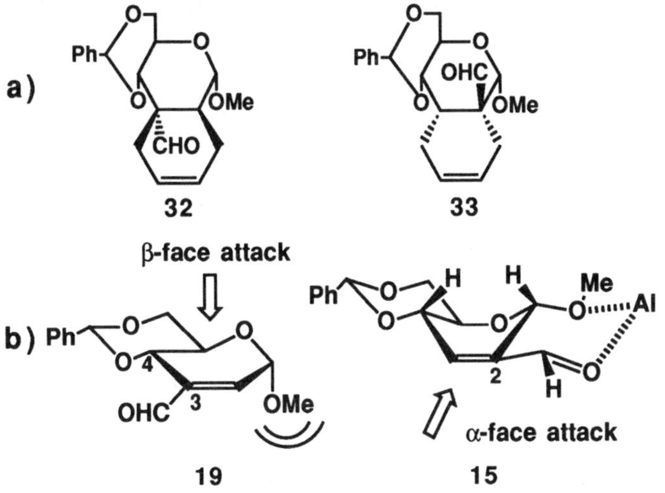

Figure 10

Homologated α-Enal Pyranosides as Heterodienes. An α,β-unsaturated aldehyde can also behave as an 1-oxabutadiene system (*25*). The resulting compound will be an "homologated pyranoside" (*26*) (term coined by Fraser-Reid (*27*) to describe two linked pyranoside units) and such compounds are useful synthons for multichiral arrays (*28*).

When the 3,2-enal **15**, was treated in refluxing ethyl vinyl ether (EVE) in the presence of $Eu(fod)_3$, the catalyst recommended by Danishefsky (*29*) to facilitate the 4π participation of α,β-unsaturated carbonyl compounds in cycloaddition reactions with electron rich olefins, the reaction took place smoothly to afford one major isomer **34** (Figure 11) in 77% yield. The adduct is the resulting of a β-face approach and an *endo* mode of addition (see A in Figure 11). Reaction of regioisomeric 2,3-enal **19** under analogous experimental conditions allowed us to isolate the homologated pyranoside **35** in 71%. The stereochemical outcome of the addition was found to be the same as for **15** (see B in Figure 11).

We thought it would be of interest to apply the same reactions to the enals **21** and **24** belonging to the *galacto*-series (*16*). 2,3-Enal **21** reacted with EVE to give a major isomer **36** in 65% yield which also resulted in a β-facial approach with *endo* selectivity. The regioisomeric 3,2-enal **24** reacted to give a single *endo* adduct **37** (68% yield) but *this time the attack took place from the α-face*, and we will put forward some explanation for this behavior at the end of the chapter.

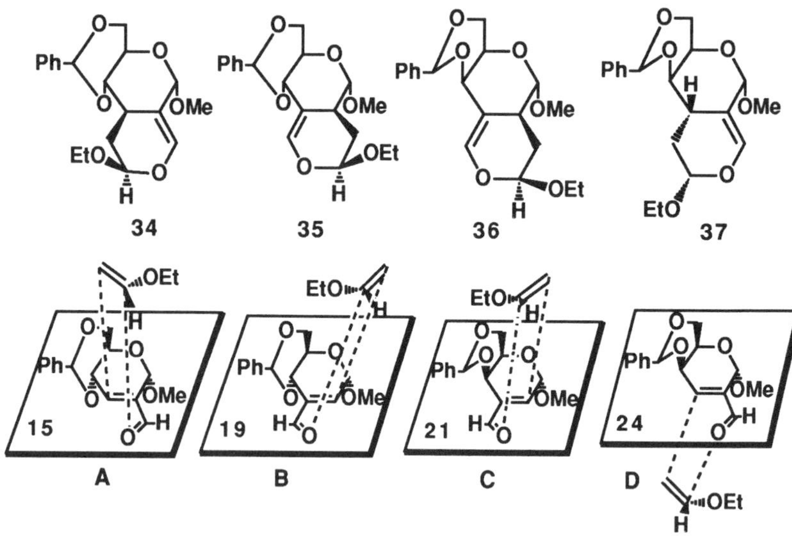

Figure 11

Cycloaddition Reactions of Pyranosido-Dienes

Prior to this study there were two precedents in the literature of cycloaddition reaction of carbohydrate derived dienes where the diene moiety was incorporated into the ring. The first example came from Fraser-Reid's laboratories (*30*) on a furanoid derived diene **38** in Figure 12, and shortly after Giuliano and Buzby reported on a pyranose diene **39** (*17*). Of special interest to us was that in both reports just one single adduct was obtained: resulting from π-facial selectivity and an exclusive *endo* mode of addition. Simultaneously to our preliminary communication Lipshutz and coworkers reported on some analogous dieno-pyranosides (*31*). More recently Giuliano has disclosed in full the work of his group on dieno-pyranosides (*32*) showing that the anomeric substituent has a strong influence in the stereochemical outcome of the cycloaddition reaction.

Results. Regioisomeric dienes **2** and **3** underwent smooth cycloaddition reactions with *sp*^{2} and *sp* type dienophiles to give annulated pyranosides under thermal conditions in refluxing toluene. Under similar conditions the dieno-furano and -pyranosides reported by Fraser-Reid's and Giuliano's groups (**38** and **39**) failed to react with *sp* dienophiles although they are in principle more reactive 1-alkoxy-1,3-diene systems (*33*).

The 2,3-diene **2** reacted with maleic anhydride to give a single stereoadduct, **40**, result of a complete π-facial selectivity through the β-face, and *endo* mode of attack. Reaction of **2** with an *sp* dienophile also took place readily in refluxing toluene to afford one single adduct, **41**, also coming from a β-face approach.

The 3,2-diene, **3**, behaved analogously and reacted with *sp* or *sp*^{2} dienophiles with complete π-facial and *endo* selectivity to give polycyclic compounds **42** and **43** respectively. The isolated dienic adduct **42** with the methoxy anomeric group missing (due to a 1,4 elimination of methanol) was the only compound that could be detected in the crude reaction mixture.

Stereoselectivity. The configuration at C-2 in compounds **40** and **41**, and at C-3 in **42** and **43** was unequivocally assigned on the basis of their vicinal coupling constants in ^{1}H NMR spectra and has been discussed elsewhere (*8*).

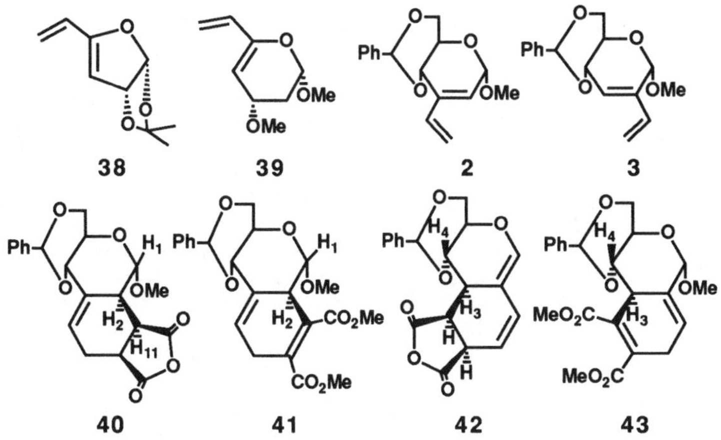

Figure 12

The *endo* mode of attack was not surprising and kept with precedents (*17, 30*). On the other hand the complete π-facial selectivity was in agreement with the previous examples in dieno-pyrano and furanosides, in every case the direction of approach was *anti* to the allylic alkoxy group and we advanced (*8*) that electrostatic interactions (*34*) were overridden in these cyclic sugar cases by the steric influence of the axial alkoxy groups in the transition state for *syn* addition.

Removal of the allylic alkoxy group in the 2,3-diene, **2**, was effected by treatment with sodium cyanoborohydride and dry hydrochloric acid in ethanol (*35*) [a similar process has recently been reported by Chapleur (*36*)] to afford compound **44** (Figure 13). Compound **44** underwent cycloaddition reaction with dimethyl acetylenedicarboxylate (DMAD) in refluxing toluene to give a mixture of two epimeric cycloadducts **45** and **46** in a 4:1 ratio.

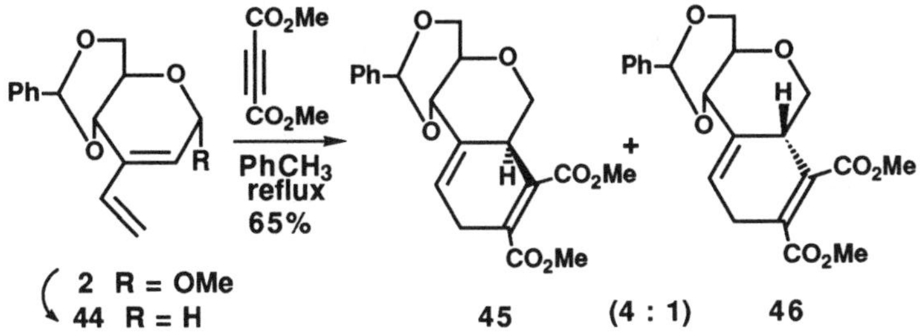

Figure 13

The loss of selectivity observed made clear the influence of the anomeric alkoxy group, but also showed that a strong stereochemical bias was inherent in the molecule, responsible for the high facial selectivity. The reason for that stereochemical preference could be ascribed to the equatorial substituent at C-5 (as advanced by Prof. Franck) fixing a preferred $^4C_1$ conformation, or possibly to the presence of the ring oxygen with two diastereotopic lone pairs.

Facial Selectivity in Diels-Alder Reactions of Dienes Containing Allylic Stereogenic Centers

All these results came to light in the midst of our ongoing interest in the π-facial selectivity in Diels-Alder reactions of homochiral dienes bearing an allylic stereogenic center. The dienes subjected to study are classified according to Kaila, Franck and Dannenberg (*37*) and are shown in Figure 14. From these studies it is clear: *a)* that the factors responsible for the observed facial stereoselection are still not fully understood; *b)* that the nature of the allylic heteroatom can have a profound effect on the course of the Diels-Alder reaction; *c)* that the geometry of the dienophile is of great importance; *d)* and that it would be risky to extrapolate the results obtained in one kind of diene to other types.

Studies in acyclic alkoxy dienes were pioneered by Franck's group (*38*) (Figure 14, X=OR), and further reports have came from McDougal (*39*) and Reitz (*40*) on acyclic dienes bearing an allylic alkoxy substituent. More recent studies in the laboratories of Kozikowski on nitrogen substituted dienes (X=N, Figure 14) (*41*), Fleming (silicon, X=Si) (*42*), and McDougal (sulfur X=S) (*43*), have made it clear that the stereochemical outcome of the cycloaddition in acyclic dienes is highly dependent on the nature of the heteroatom at the allylic center of the diene. Macaulay and Fallis (*44*) found a striking reversal of facial selectivity in cyclic dienes when changing from oxygen to methyl-sulphide (X=O to X=S in Figure 14), and Kato and coworkers (*45*) reported some differences in behavior for sulfur and selenium substituted cyclic dienes (X=S; X=Se in Figure 14).

The geometry of the dienophile can exert a significant control on the diastereofacial selectivity of acyclic dienes having allylic substituents in Diels-Alder reactions, as has been shown by Kozikowski (*41*) and Franck (*46*)

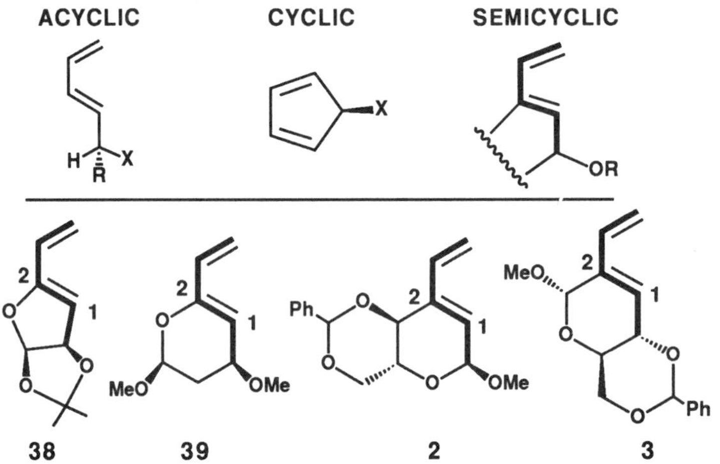

Figure 14

Carbohydrate-Derived Dienes as a Model for Semicyclic Dienes. We thought that carbohydrate-derived dienes could be employed as model systems for semicyclic dienes as shown in Figure 14, (although there can be some exceptions as recently reported by Giuliano, that has shown how the stereodirecting effect of an alkoxy allylic substituent can be overridden by the effect of the anomeric substituent (*32*)). The π-facial selectivity in dienes, **2**, **3** and **38**, **39**, *anti* to the allylic alkoxy group, is the same as the one found by Overman and Hehre in semicyclic dienes (*47*).

Semicyclic Dienes with One or Two Allylic Stereogenic Centers. The semicyclic dienes studied by Fraser-Reid **38** and Giuliano **39** (see Figure 14) bear just one single allylic stereogenic center at position 1 (diene numbering) and the addition took place from the face opposite the allylic alkoxy group.

On the other hand the dieno-pyranosides studied by us (*8*) and Lipshutz (*31*) have two allylic stereogenic centers (positions 1 and 2 in dienes **2** and **3**, Figure 14), with the same orientation so that reinforcement in the *anti* stereodirecting effect could be predicted.

Semicyclic Dienes with One Allylic Stereogenic Substituent at Position 2 in the Diene Moiety. From the above mentioned reasoning, the question arose about if any of the stereodirecting effect would be motivated by the additional allylic substituent at position 2 and, furthermore, if a single allylic substituent at position 2 in the diene system (C in Figure 15) would have some effect in the stereochemical outcome of the cycloaddition reaction.

Choice of the Model Substrates. In this context, we decided to use of 3-substituted C-2 vinyl glycals (Figure 15) as model compounds for 2-substituted dienes bearing a chiral centre in the 2-substituent (C). Prior studies have addressed 1-substituted dienes carrying an allylic chiral centre on the 1-substituent (A in Figure 15).

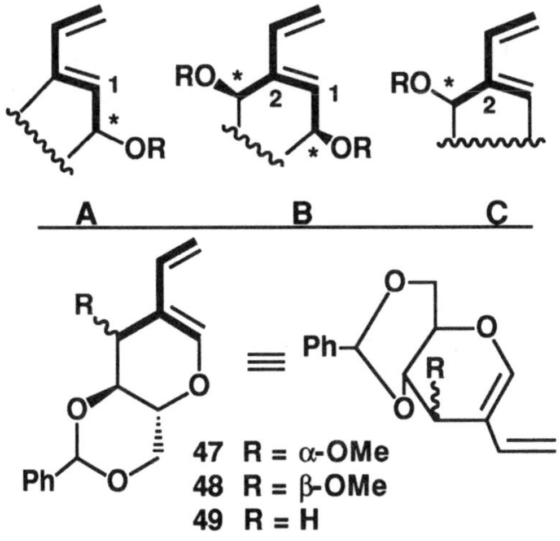

A B C

47 R = α-OMe
48 R = β-OMe
49 R = H

3-Substituted C-2 Vinyl Glycal

Figure 15

The preparation of the glycals **47-48** was uneventfully carried out following the methods developed by us. Diene **49**, without any substituent at C-3 was chosen for us to have a blank reference on the stereochemical bias inherent to the carbohydrate moiety *vide supra* (Figure 16).

Results. When glycals **47-48**, as shown in Figure 16, were subjected to reaction with *sp* dienophiles, in every instance, a mixture of epimeric tricyclic compounds, **50**, resulting from attack at both faces were obtained. This implies that the allylic center at position 2 does not have any important stereodirecting effect when *sp* dienophiles were used [along with adducts **50**, some aromatic compounds (*48, 49*) could also be isolated (*50*)].

Unsubstituted glycal **49** underwent cycloaddition reaction with a sp^2 dienophile (maleic anhydride) in refluxing toluene to give a mixture of two epimeric compounds arising from an expected *endo* attack. The 7:3 ratio in stereoadducts **51** and **52** clearly indicates an stereochemical preference inherent to the pyranosidic substrate (*vide supra*). This predicts that, the presence of an α-oriented alkoxy group as in **47**, would reinforce the natural preference of the substrate for a β-face attack, and this is experimentally observed. Reaction of **47** with maleic anhydride furnished one single "annulated C-glycopyranoside" **53** (*50*).

When the directing group is β-oriented, as in **48** (Figure 16) cycloaddition reaction with maleic anhydride took place to give one single adduct, **54**, resulting from an α-facial approach.

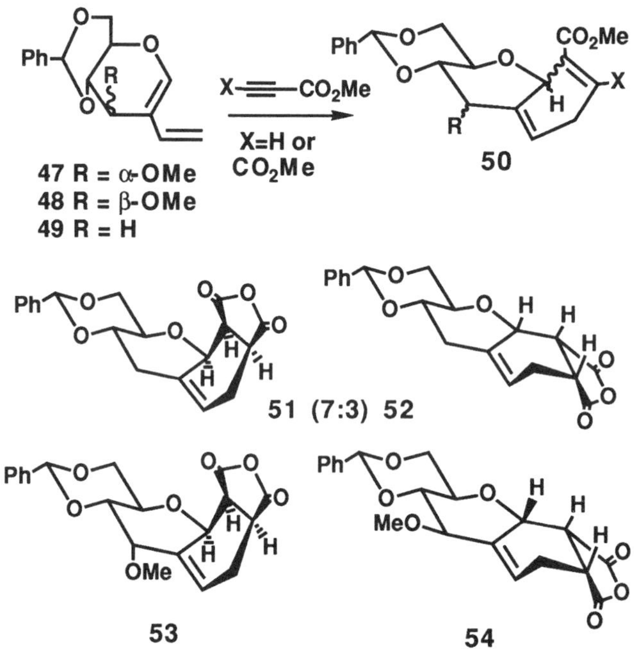

X—≡—CO_2Me

X=H or CO_2Me

47 R = α-OMe
48 R = β-OMe
49 R = H

50

51 (7:3) **52**

53

54

Figure 16

Discussion. The experimental results clearly indicate that, for 2-substituted dienes bearing a chiral centre in the 2-substituent, the nature of the dienophile has a drastic effect in the relative topicity of the Diels-Alder reaction. Our rationalization for such behavior is shown in Figure 17 where a more significant interaction with the allylic substituent would be expected for the *sp*² dienophile, through an *endo* transition state in a hypothetical syn approach.

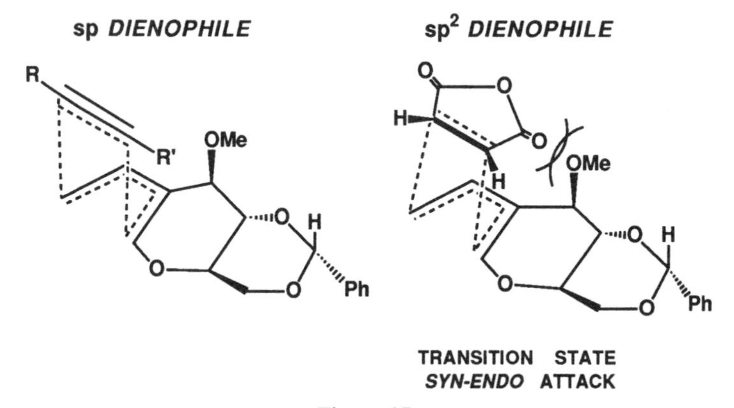

Figure 17

Stereodirecting Effect of Two Allylic Oxygenated Substituents

The results presented seemed to indicate that carbohydrate derived dienes with just one allylic substituent at position 1 in the dienic system (**38** and **39**, Figure 14, A in Figure 15) exert an *anti* directing effect and that is in agreement with the result by Overman and Hehre in semicyclic dienes (*47*). One single allylic substituent at position 2 in the dienic system (as in **39** and **40** in Figure 14, B in Figure 15) exert an *anti* directing effect when sp^2 dienophiles are utilized as counterparts. By corollary, two allylic stereogenic centers at positions 1 and 2 in a *syn* relationship (C in Figure 15) in the dienic system would reinforce an *anti* directing effect as we have shown.
In this context, the question about the stereodirecting properties of two allylic stereogenic centers in an *anti* relationship seemed of interest and carbohydrates, because of their highly oxygenated nature, appeared to be the substrates of choice. Pyranosido-dienes derived from α-methyl galactose (**26** and **27**) would fulfill such a requisite.

Stereoselectivity in Hetero Diels-Alder Reactions of α-Enals. Although we have not carried out any experiment in pyranosido-dienes of the *galacto*-series, some of our results in cycloaddition reactions of heterodiene pyranosides (Figure 11) can now be rationalized. Stereoadducts **34** and **35** are formed from enals **15** and **19** respectively, through a β-face approach, i.e., *anti* to both allylic alkoxy substituents (that are in a *syn* relationship). Enals **21** and **24** that possessing two allylic alkoxy substituents in an *anti* relationship react through opposite faces (compare C and D in Figure 11) to give homologated pyranosides **36** and **37**, which seems to indicate that the effect on the relative topicity in the reaction is ruled by the allylic substituent at position 1 in a 1,2 disubstituted diene.

Acknowledgments. The authors would like to express their gratitude to all the people involved in this project, and whose names appeared in the references. Special thanks are given to graduate students Eric Lameignère and Catherine Burnouf and to Drs Alberto A. Ghini and María. A. Laborde for their enthusiasm. We also thank Dr. A. Olesker for his continuous support.

Literature Cited

1. Diels, O.; Alder, K. *Liebigs Ann. Chem.* **1928**, *460*, 98.
2. (a) Paquette, L. A. *Asymmetric Synthesis*; Morrison, J. D., Ed.; Academic Press: New York, 1984; Chapter 7. (b) Desimoni, G.; Tacconi, G.; Bario, A.; Pollini, G. P. *Natural Products Syntheses through Pericyclic Reactions*; ACS Monograph 180; American Chemical Society: Washington, DC, 1984, Chapter 5. (c) Oppolzer, W. *Angew. Chem., Int. Ed. Engl.* 1984, 23, 876. (d) Ciganek, E. *Org. React.* (*N.Y.*) **1984**, 32,1. (e) Gleiter, R.; Böhm, M. C. *Pure Appl. Chem.* **1983**, *55*, 237.
3. (a) Fraser-Reid, B.; Anderson, R. C. in *Fortschr. Chem. Org. Naturst.* **1980**, *30*, 1. (b) Hannessian, S. In *Total Synthesis of Natural Products: The 'Chiron' Approach;* Baldwin, J. E., Ed.; Pergamon: Oxford, 1983.
4. Fraser-Reid, B.; Tsang, R. *Strategies and Tactics in Organic Synthesis*; Lindberg T., Ed.; Academic Press: New York, NY, 1989; Vol. 2.
5. a) Fitzsimmons, B. J.; Fraser-Reid, B. *J. Am. Chem. Soc.* **1979**, *101*, 6123. b) Fitzsimmons, B. J.; Fraser-Reid, B., *Tetrahedron*, **1984**, 1279
6. Bonnert, R. V.; Jenkins, P. R. *Tetrahedron Lett.* **1987**, *28*, 697
7. Baker, D. C.; Brown, D. K.; Horton, D.; Nickol, R. G. *Carbohydrate Res.* **1974**, *32*, 299. Horton, D.; Just, E. K. *Carbohydrate Res.* **1971**, *18*, 81.
8. López, J. C.; Lameignère, E.; Lukacs, G. *J. Chem. Soc., Chem. Commun.* **1988**, 706.

9. *Synthetic Reagents*; Pizey, S. S., John Wiley & Sons Ed.; New York, NY, 1974; Vol 1. pp 321-357.
10. Lagrange, A.; Olesker, A.; Lukacs, G.; Thang, T. T. Carbohydr. Res. **1982**, *110*, 165.
11. López, J. C.; Lameignere, E.; Burnouf, C.; Laborde, M. A.; Ghini, A. A.; Olesker, A.; Lukacs, G. manuscript in preparation.
12. Corey, E. J.; Seebach, D. *Angew. Chem., Int. Ed.* **1965**, *4*, 1075.
13. Sepulchre, A. M.; Lukacs, G.; Vass, G.; Gero, S. D. *C. R. Acad. Sc. Serie C*, **1971**, 1180. b) Sepulchre, A. M.; Lukacs, G.; Vass, G.; Gero, S. D., *Angew. Chem., Int. Ed.* **1972**, *11*, 148. c) Sepulchre, A. M.; Lukacs, G.; Vass, G.; Gero, S. D., Bull. Soc. Chim. Fr. **1972**, 4000.
14. Gröbel, B.-T, Seebach, D. *Synthesis* **1977**, 357.
15. López, J. C.; Lameignère, E.; Lukacs, G. *J. Chem. Soc., Chem. Commun.* **1988**, 514.
16. Lameignère, E. Ph. D. Dissertation Universitè de Paris-Sud, 1990.
17. Giuliano, R. M.; Buzby, J. H. *Carbohydr. Res.* **1986**, *158*, c1.
18. Buchanan, J. G.; Sable, H. Z. In *Selective Organic Transformations* Thyagarajan, B. S. Ed. Wiley-Interscience, New York, NY, 1972, Vol. 2; pp 1-95.
19. Asano, T.; Yokota, S.; Mitsunobu, O. *Chem. Lett.* **1983**, 343.
20. Burnouf, C.; López, J. C.; Laborde, M. A.; Olesker, A.; Lukacs, G. *Tetrahedron Lett.* **1988**, *29*, 5533.
21. Earnshaw, C.; Wallis, C. J.; Warren, S. *J. Chem. Soc., Perkin Trans. 1.* **1979**, 3099.
22. Primeau, J. L.; Anderson, R. C.; Fraser-Reid, B. *J. Chem. Soc., Chem. Commun.* **1980**, 6.
23. Gnichtel, H.; Gumprecht, C.; Luger, P. *Liebigs Ann. Chem.* **1984**, 531.
24. Martin, S. F. *Tetrahedron* **1980**, 419.
25. a) Desimoni, G.; Tacconi, G. *Chem. Rev.* **1975**, *75*, 651.b) Tiezte, L. F. In *Selectivity- a Goal for Synthetic Efficiency*, Trost, B. M., Bartmann, W., Eds.; Verlag Chemie, Weinheim, 1984, pp 299-315. c) Schmidt, R. R. *Pure Appl. Chem.* **1987**, *59*, 415. d) Boger, D. L.; Weinreb, S. M. Hetero Diels-Alder Methodology in Organic Synthesis; Academic: New York, N. Y. 1987.
26. Fraser-Reid, B.; Magdzinski, L.; Molino, B. *J. Am. Chem. Soc.* **1984**, *106*, 731.
27. a) Molino, B. F.; Magdzinski, L.; Fraser-Reid, B. *Tetrahedron Lett.* **1983**, *24*, 5819. b) Magdzinski, L.; Cweiber, B.; Fraser-Reid, B. *Tetrahedron Lett.* **1983**, *24*, 5823.
28. Chapleur, Y.; Euvrard, M. N. *J. Chem. Soc., Chem. Commun.* **1987**, 884.
29. Bernardsky, M.; Danishefsky, S. *J. Am. Chem. Soc.* **1983**, *105*, 3716.
30. Sun, K. M.; Fraser-Reid, B.; Tam, T. F.*J. Am. Chem. Soc.* **1982**, *104*, 367. Sun, K. M.; Giuliano, R. M.; Fraser-Reid, B. *J. Org. Chem.* **1985**, *50*, 4774.
31. Lipshutz, B. H.; Nguyen, S. L.; Elworthy, T. R. Tetrahedron **1988**, *44*, 3355.
32. Giuliano, R. M.; Buzby, J. H.; Marcopulos, N.; Springer, J. P. *J. Org. Chem.* **1990**, *55*, 3555. Coleman, R. S. *Chemtracs-Org. Chem.* **1991**, *4*, 60.
33. Fringuelli, F.; Taticchi, A.; *Dienes in the Diels-Alder Reaction;* John Wiley & Sons: New York, N. Y., 1990.
34. Kahn, S. D.; Hehre, W. J. *J. Am. Chem. Soc.* **1987**, *109*, 663.
35. Ghini, A. A. unpublished results.
36. Chapleur, Y.; Boquel, P.; Chrètien, F. *J. Chem. Soc. Perkin Trans. I* **1989**, 703.
37. Kaila, N.; Franck, R. W.; Dannenberg, J. J. *J. Org. Chem.* **1989**, *54*, 4206.
38. Franck, R. W.; Argade, S.; Subramanian, C. S.; Frechet, D. M. *Tetrahedron Lett.* **1985**, *27*, 3187.
39. McDougal, P. G.; Rico, J. G.; VanDerveer, D. *J. Org. Chem.* **1986**, *51*, 4492.
40. Reitz, A. B.; Jordan, A. D. Jr.; Maryanoff, B. E. *J. Org. Chem.* **1987**, *52*, 4800.

41. Kozikowski, A. P.; Nieduzak, R. R.; Konoike, T.; Springer, J. P. *J. Am. Chem. Soc.* **1987**, *109*, 5167. Kozikowski, A. P.; Jung, S. H.; Springer, J. P.*J. Chem. Soc., Chem. Commun.* **1988**, 167.
42. Fleming, I. *Pure Appl. Chem.* **1988**, *60*, 71.
43. McDougal, P. G.; Jump, J. M.; Rojas, C.; Rico, J. G. *Tetrahedron Lett.* **1989**, *30*, 3897.
44. Macaulay, J. B.; Fallis, A. G. *J. Am. Chem. Soc.* **1988**, *110*, 4074.
45. Ishida, M.; Aoyama, T.; Kato, S. *Chem. Lett.* **1989**, 663.
46. Tripathy, R.; Franck, R. W.; Onan, K. D. *J. Am. Chem. Soc.* **1988**, *110*, 3257.
47. Fisher, M. J.; Hehre, W. J.; Kahn, S. D.; Overman, L. E. *J. Am. Chem. Soc.* **1988**, *110*, 4625.
48. Hecker, S. J.; Heathcock, C. H. *J. Org. Chem.* **1985**, *50*, 5159.
49. Gnichtel, H.; Gumprecht, C. *Liebigs Ann. Chem.* **1985**, 628.
50. Burnouf, C.; López, J. C.; Garcia Calvo-Flores, F.; Laborde, M. A.; Olesker, A.; Lukacs, G. *J. Chem. Soc., Chem. Commun.* **1990**, 823.

RECEIVED December 26, 1991

Chapter 4

Synthesis of β-Lactams from Unsaturated Sugars and Isocyanates

Marek Chmielewski, Zbigniew Kałuża, Jacek Grodner, and Romuald Urbański

Institute of Organic Chemistry, Polish Academy of Sciences, 01–224 Warsaw, Poland

Cycloaddition of trichloroacetyl isocyanate to glycals proceeds satisfactorily at room temperature to give a mixture of [2+2] and [4+2]cycloadducts. The isocyanate attacks the glycal stereospecifically *anti* with respect to the C-3 substituent. Glycolic cleavage of the *vic*-diol present in 2-C:1-N-carbonyl-2-deoxy-glycopyranosylamines with sodium meta-periodate provides useful starting materials for the synthesis of 1-oxabicyclic β-lactams.

Owing to the importance of β-lactam antibiotics, the [2+2]cycloadditions of ketenes to imines and isocyanates to olefins play a special role because both reactions lead to direct formation of a four-membered azetidinone-2 ring. The first one can be performed using a large variety of ketenes and ketene equivalents,[1,2] whereas the second one leads to useful compounds only in the case of chlorosulfonyl isocyanate and vinyl acetates,[3] vinyl silyl ethers,[4] and dienes.[5] One example of the addition of an isocyanate to a vinyl ether has been reported (Scheme 1).[6]

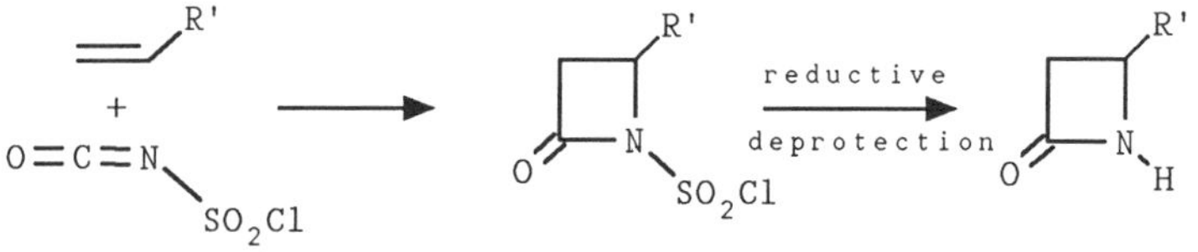

R' = OAc, CH=CHOAc, $CH=CH_2$

Scheme 1

0097–6156/92/0494–0050$06.00/0

It is known that nucleophilic olefins, such as vinyl ethers, acetates or enamines, and highly electrophilic sulfonyl or acyl isocyanates, give cycloadducts in good yields, but only in a few cases are the adducts stable enough to be isolated.[7] The electron-donating substituent (X) in the olefin and the electron-withdrawing substituent (Z) in the isocyanate, necessary for cycloaddition to occur, are responsible also for the low stability of the cycloadduct, inducing heterolytic cleavage of the N–C-4 bond and formation of α,β-unsaturated amides *via* a zwitterionic intermediate (Scheme 2). Therefore, in order to obtain a stable structure, it is necessary to remove the electron-withdrawing substituent from the nitrogen atom prior to purification or to any other transformation of the [2+2]cycloadduct. This, so far, generally has been accomplished successfully in the case of the chlorosulfonyl group.[7]

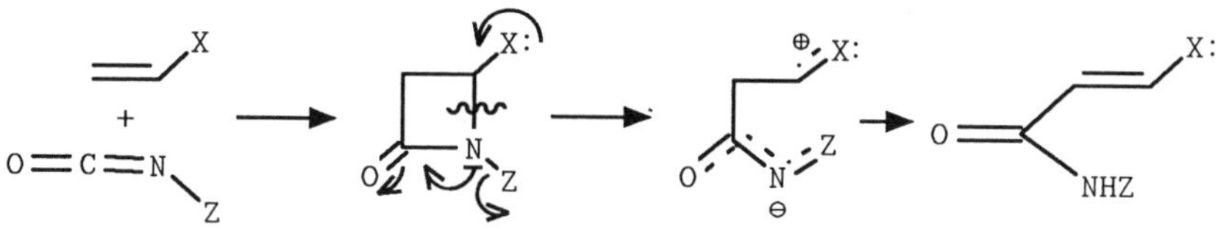

X: = OAcyl, OAlkyl, NR^1R^2

Z = SO_2R^3, Acyl

Scheme 2

Several years ago, we initiated a synthetic project leading from sugars to 1-oxabicyclic β-lactams (Scheme 3). This has prompted us to investigate the [2+2]cycloaddition of isocyanates to glycals and to the related dihydro-2H-pyran derivatives. The direct formation of a β-lactam ring was the crucial step in the planned synthesis.

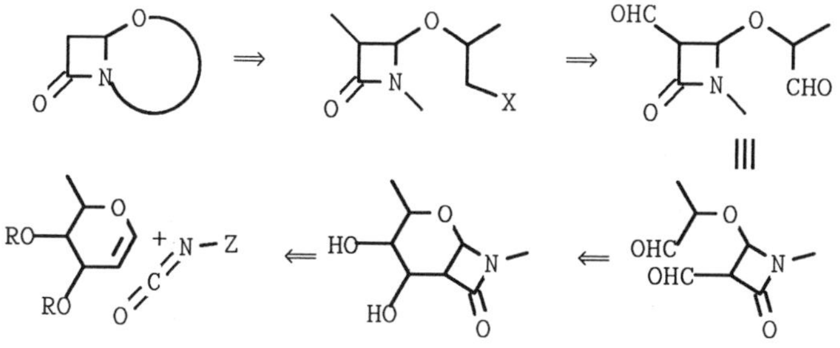

Scheme 3

The reaction between tri-O-acetyl-D-glucal 7 and chlorosulfonyl isocyanate has been studied in the past, but neither formation of a cycloadduct nor of a rearranged product has been observed.[8] Isocyanate acted only as acid catalyst causing decomposition of sugar material. On the other hand, [2+2]cycloaddition of active isocyanates to dihydro-2H-pyran and to its derivatives has been widely investigated,[9,10,11,12] under a variety of conditions. The reaction of tosyl isocyanate with dihydro-2H-pyran **1** at low temperature (0°) led to the formation of bicyclic β-lactam **2**.[9] Elevation of the cyclization temperature resulted in the rearrangement of the four-membered ring to the open-chain amide **3** (Scheme 4).

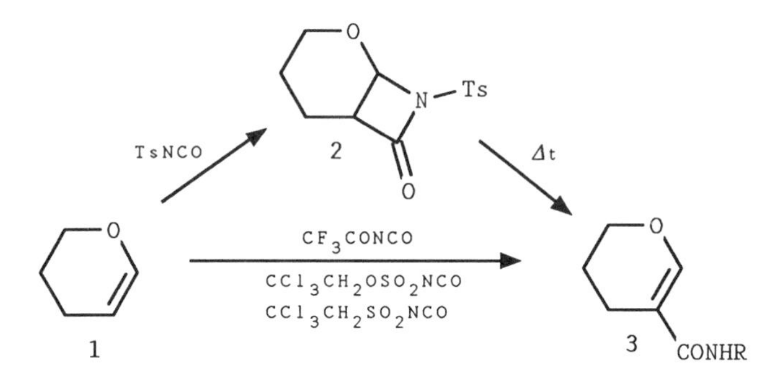

R=Ts, $CCl_3CH_2OSO_2$, $CCl_3CH_2SO_2$, CF_3CO

Scheme 4

Recently Barrett at al.[10] found that 2,2,2-trichloroethylsulfonyl, 2,2,2-trichloroethoxysulfonyl, and trifluoroacetyl isocyanate reacted with **1** to give the respective unsaturated amide **3**, no β-lactam being isolated in all three cases. Similar results were reported by Chan and Hall[11] for monosubstituted dihydro-2H-pyrans.

CCl_3CONCO, $CHCl_3$ or CH_3CN; $COCCl_3$; CCl_3; $CONHCOCCl_3$

1 4 5 6

Scheme 5

Chitwood, Gott, and Martin[12] found that when **1** was treated with trichloroacetyl isocyanate, the unsaturated amide **6** was obtained *via* the unstable β-lactam **4** and the [4+2]cycloadduct **5** (Scheme 5). They observed that the reaction performed in acetonitrile proceeded ten times faster than in chloroform. Owing to their low stability, both by-products **4** and **5** were detected only by NMR.

Inferences drawn from literature data[8-12] prompted us to apply the high-pressure technique to the [2+2]cycloaddition of isocyanates to glycals. The lack of reactivity at room temperature and atmospheric pressure, the exceptional sensitivity of cycloadducts to temperature, and the large negative activation volume of [2+2]cycloadditions,[13] in contrast to the usually low sensitivity of intramolecular rearrangements to this parameter,[14] strongly supported the use of high pressure for this reaction.

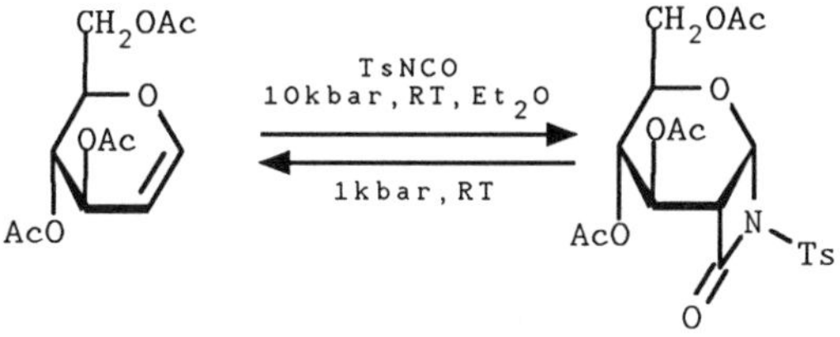

CH_2OAc O OAc AcO — TsNCO, 10kbar, RT, Et_2O ⇌ 1kbar, RT — CH_2OAc O OAc AcO N Ts O

7 8

Scheme 6

Application of pressures of 10-11 Kbar greatly facilitates the [2+2]cycloaddition of tosyl isocyanate to acetylated glycals[15,16] (Scheme 6). The reactions were performed overnight in ethyl ether at room temperature or at 50 °C. Usually a crystalline product was obtained. The reaction showed high stereoselectivity, affording products with the four-membered ring *anti* with respect to the acetoxy group at C-3. Upon heating, or even at room temperature, the adducts (Scheme 6) underwent retro addition to afford the glycal.[16,17] The rate of retro addition increased with a rise of temperature and polarity of the solvent. This unexpected tendency of cycloadducts for retro addition explains why β-lactams could not be obtained from glycals and isocyanates under thermal conditions.[8,17] High pressure not only accelerates the reaction rate but also affords cycloadducts that are thermodynamically unstable at normal pressure. Moreover, demonstration, for the first time, of the reversibility of the isocyanate cycloaddition to olefins puts a new light on the mechanism of this reaction and on the potential for the preparation of cycloadducts under conditions of normal pressure. The diastereoselectivity of the [2+2]cycloaddition of tosyl isocyanate to 3-deoxyglycals was relatively low.[16] The isocyanate entered preferentially *anti* with respect to the terminal C-6 carbon atom.

The reaction of glucal **7** with 2,2,2-trichloroethoxysulfonyl isocyanate under 10 Kbar pressure at room temperature in ether solution afforded a mixture of β-lactam **9**, unsaturated amide **10** and unreacted substrate (Scheme 7).[18] The contents of **7**, cycloadduct **9**, and amide **10** depended on the reaction time, as shown. When treated with alcohols or water at room temperature, all β-lactam-adducts underwent a rapid opening of the four-membered ring to give the respective glycosides or cyclic hemiacetals (Scheme 8).[16,18] The opening of the β-lactam ring proceeded stereospecifically with inversion of the configuration at the C-1 carbon atom.

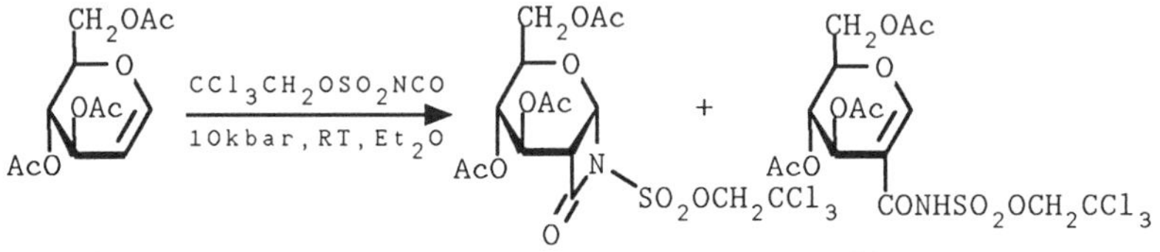

$CCl_3CH_2OSO_2NCO$, 10kbar, RT, Et_2O

	7	9	10
1h	27.3%	54.5%	18.2%
5h	17.2%	48.3%	34.5%
8h	16.7%	34.9%	48.7%

Scheme 7

The reversibility and thermodynamic control of product formation found for the high-pressure reaction between glycals and tosyl isocyanate[16] indicated that the [2+2]cycloaddition of isocyanates to glycals could occur at atmospheric pressure under specific reaction conditions including an excess of isocyanate, as well as proper selection of solvent and substrates. Acyl isocyanates are generally less reactive in [2+2]cycloaddition reactions than sulfonyl isocyanates, except for trichloro- and trifluoroacetyl isocyanate.[10,12] In addition, acyl isocyanates are problematic because of the competitive formation of [4+2]cycloadducts, which are usually thermodynamically preferred over the [2+2]cycloadducts.[12]

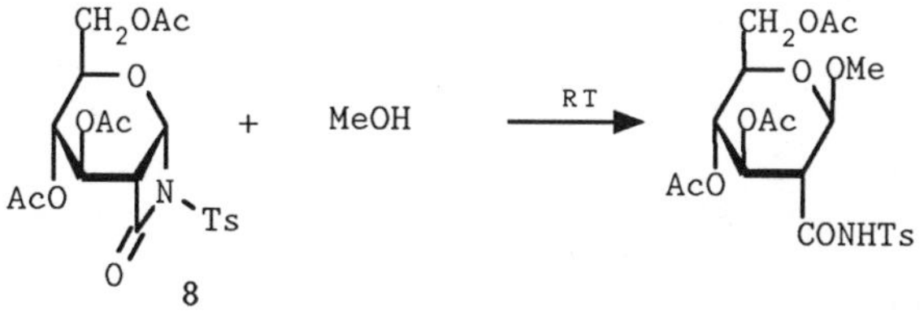

Scheme 8

The reaction between di-O-acetyl-L-rhamnal **11** and trichloroacetyl isocyanate was investigated under various conditions, under high and atmospheric pressure. The results confirmed the expectations based on the previous experiments using tosyl isocyanate and acetylated glycals under high pressure.[15,16] The isocyanate approached rhamnal **11** preferentially *anti* with respect to the 3-O-acetyl group (Scheme 9). The percentages of the substrate and products as a function of the solvent, pressure, temperature, and initial proportion of substrates were studied.[18] The formation of substantial amounts of the β-*manno* cycloadduct **14** is noteworthy, and can be attributed to the complexation of the isocyanate by the 3-O-acetyl group.

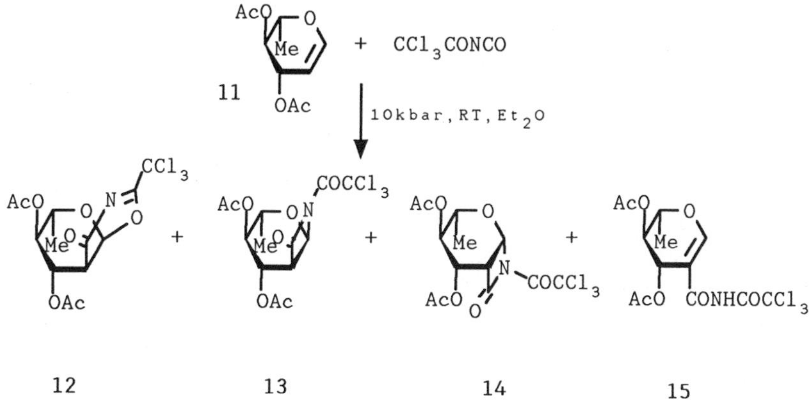

Scheme 9

Consistent with reported data, free energies of activation of the cycloaddition, of the reverse reaction, and of the rearrangement to the α,β-unsaturated amide, do not differ considerably. Hence, a slight modification of the glycal moiety, or even a change in the solvent could shift the reaction towards the desired cycloadduct, or subsequently towards the unsaturated amide. Polar solvents, such as acetonitrile, acted in a dual manner, causing acceleration of the reaction rate and a shift of the equilibrium towards the cycloadducts.

The addition of trichloroacetyl isocyanate to rhamnal **11** in ethyl ether under 10 Kbar pressure was used for the preparation of the [4+2]cycloaddition product **12** (Scheme 9).[18] During the high-pressure experiment, compound **12** crystallized in good yield and in nearly pure form . Cycloadduct **12** was obtained stereospecifically, as a result of isocyanate addition *anti* with respect to the C-3 acetoxy group. The stereospecificity of the formation of **12** could be attributed to the kinetic control of addition by the C-3 substituent; nevertheless, a shift of the equilibrium: β-**manno** [4+2] ⇄ β-**manno** [2+2]cycloadduct, entirely towards the β-lactam should also be taken into consideration. The lack of stabilizing anomeric effect of the imidate function as well as axial location of the carbonyl group at the C-2 carbon atom in the pyranoid ring of the β-manno [4+2]cycloadduct support this assumption. Attempts at isolation of cycloadducts **13** and **14** from the reaction mixture failed owing to their high reactivity. Therefore, *N*-deprotection was necessary prior to isolation or chemical transformation of **13** and **14**.[18] *N*-Unsubstituted β-lactam **16** was obtained by passing a mixture of **12** - **15** through a Florisil column. Compound **16** was accompanied by amide **17** and α,β-unsaturated amide **18**. The low stability of the β-manno adduct **14**, manifesting itself by the absence of the respective *N*-unsubstituted β-lactam, was noteworthy.[18] The mixture of **13** and **14**, obtained after separation of **12**, treated with methanol afforded a mixture of glycosides **19** and **20** with the α-L-**gluco** and β-L-**manno** configuration, respectively. Traces of β-lactam **16** and amide **18** were also isolated.[18]

The results of the high-pressure experiments aided in the selection of the most effective substrates and conditions for performing cycloadditions of isocyanates to glycals at atmospheric pressure. Trichloroacetyl isocyanate and glycals having nonpolar protecting groups were found to be the most suitable substrates for cycloaddition, providing relatively stable cycloadducts, and offering an isocyanate-activating substituent readily removable from the cycloadducts under mild conditions.[19,20,21,22]

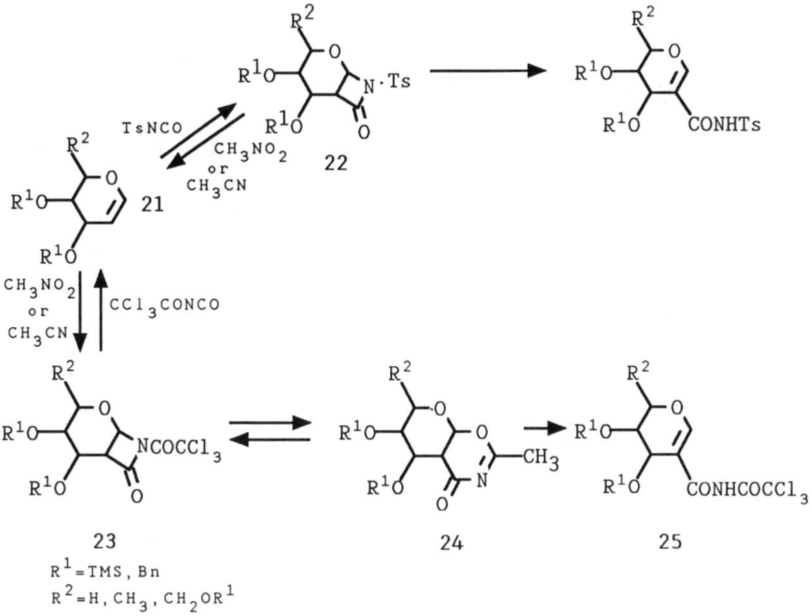

R^1=TMS, Bn
R^2=H, CH_3, CH_2OR^1

Scheme 10

Glycals **21** were found to react readily under atmospheric pressure with 2–3-molar equiv. of tosyl or trichloroacetyl isocyanate. The rate of addition and the proportion of the substrate and products differed in dependence on the solvents and substrates used. The reactions carried out under atmospheric pressure proceeded with high stereoselectivity to afford *cis*-fused bicyclic systems in which addition had taken place *anti* with respect to the C-3 substituent (Scheme 10). This high stereoselectivity is noteworthy, in contrast to that found for the cycloaddition of trichloroacetyl isocyanate to **11**.[18] Tosyl isocyanate gave the respective β-lactams **22** after 6-40 h. Subsequently slow rearrangement of the adduct into the respective α,β-unsaturated amide was observed.[21] Attempts to cleave the tosyl substituent from the nitrogen atom of **22** in order to obtain stable *N*-unsubstituted compounds, without decomposition of the β-lactam ring, were unsuccessful and this reaction was not investigated any further.

Trichloroacetyl isocyanate reacted with glycals **21** producing [2+2]cycloadducts **23**, [4+2]adducts **24** and the open-chain amide **25**. The initial proportion of the adducts **23** and **24** changed slowly, resulting in the predominance of **24**. In all cases, both bicyclic intermediates slowly rearranged to amide **25**.

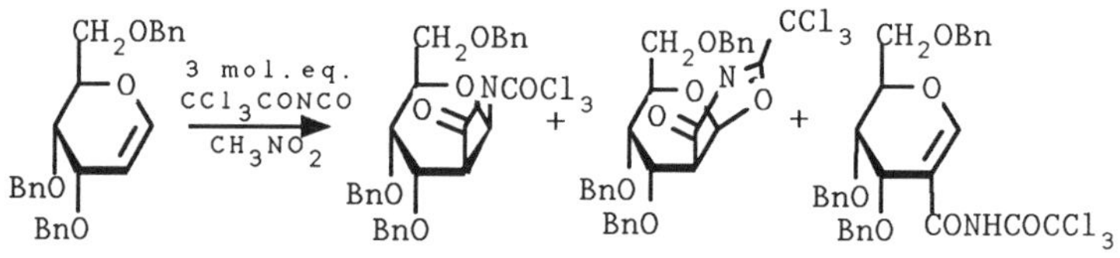

26 27

Scheme 11

Polar solvents, such as acetonitile or nitromethane accelerated the reaction rate, more so the cycloaddition than the rearrangement to the amides. An increase in polarity of the solvent promoted formation of the [2+2]cycloadduct as a kinetic product. The rate of cycloaddition depends on the concentration of both components, whereas the intramolecular rearrangement of cycloadducts to **25** does not. The highest content of [2+2]cycloadduct in the reaction mixture could be achieved when the cycloaddition was performed with more concentrated solutions of benzylated glycals in a polar solvent.[22] For example, tri-O-benzyl allal **26** (30 mg/ml) plus 3 molar equiv. of trichloroacetyl isocyanate in nitromethane produced 42% of the respective β-lactam **27** after 10 hrs, whereas at a glycal-concentration of 400 mg/ml, **27** was formed in a 60% yield after 5 hrs.[22] (Scheme 11)

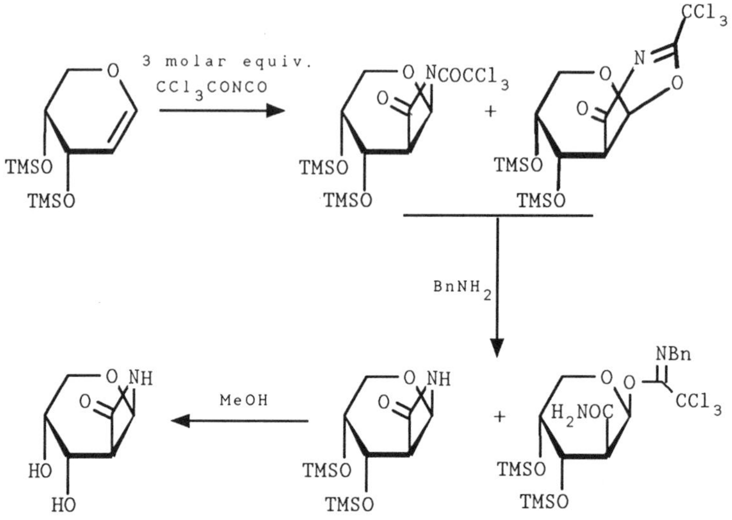

28

Scheme 12

Addition of a primary amine to a mixture of **23** and **24** quenched the reaction progress, leading to removal of *N*-protection and to formation of stable β-lactams. Further deprotection of oxygen atoms afforded stable bicyclic β-lactams.[20-23]

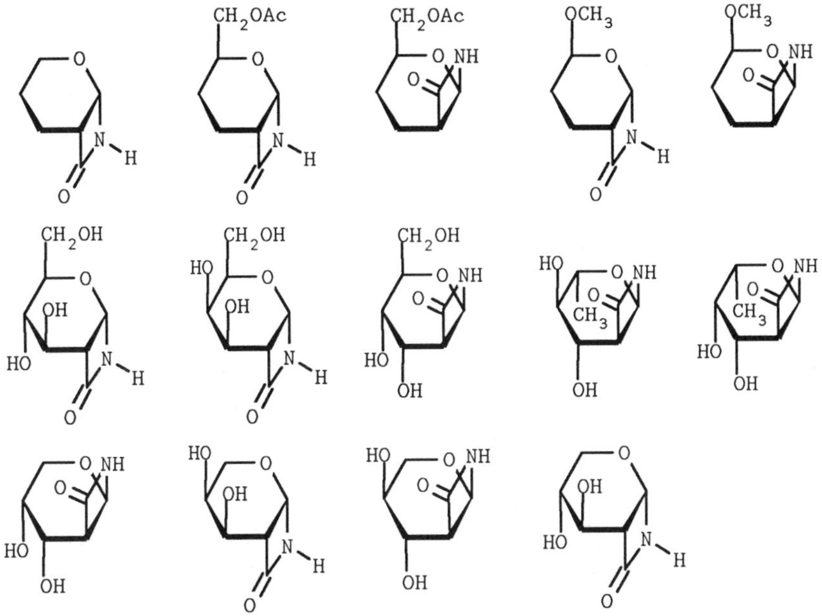

Scheme 13

This is illustrated in Scheme 12 by the example of the synthesis of β-lactam **28**.[20] Until now 14 structures derived from dihydro-2H-pyrans and glycals were obtained, and are shown in Scheme 13. The same procedure could be applied to the cycloaddition of trichloroacetyl isocyanate to the furanoid glycal **29**.[23] The reaction proceeded with the same high stereoselectivity to produce [2+2] and [4+2]cycloadducts having the α-D-gluco configuration (Scheme 14). *N*-Deprotection led to the formation of stable compound **30**.

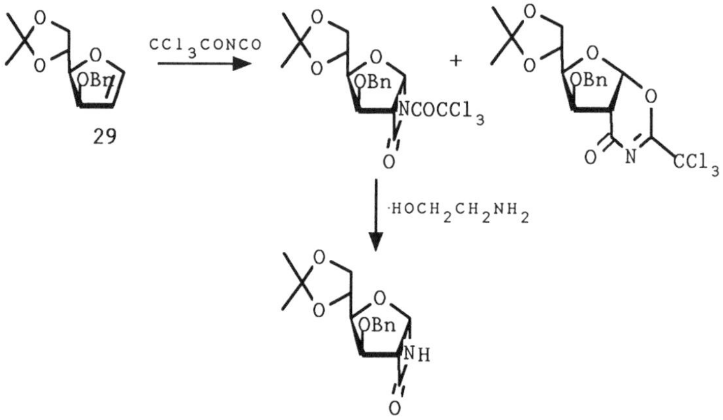

Scheme 14

Periodate oxidation of compound **28** lacking *N*-protection, under standard conditions[17], led to the formation of complex mixtures of products which were not investigated. In order to obtain more stable compounds, sugar β-lactams were benzylated at the nitrogen atom in a three-step procedure to afford compounds **31** (Scheme 15).[24] *N*-Benzyl β-lactam **31** could also be obtained using another sequence of reactions, which was found to be particularily attractive in the case of cycloadducts obtained from benzylated glycals, for which a high content of the [2+2]cycloadduct was observed.[22,24] For example, galactal **32** and trichloroacetyl isocyanate in an acetonitrile or nitromethane solution reacted to afford a mixture of cyloadducts containing up to 75% of the desired [2+2]cycloadduct (Scheme 16). Compound **33**, thus obtained, was benzylated at the nitrogen atom according to the procedure described for silylated compounds (Scheme 16), and subsequently was subjected to hydrogenolysis in the presence of a palladium catalyst affording **34**; no hydrogenolysis of the *N*-benzyl group was observed.[24]

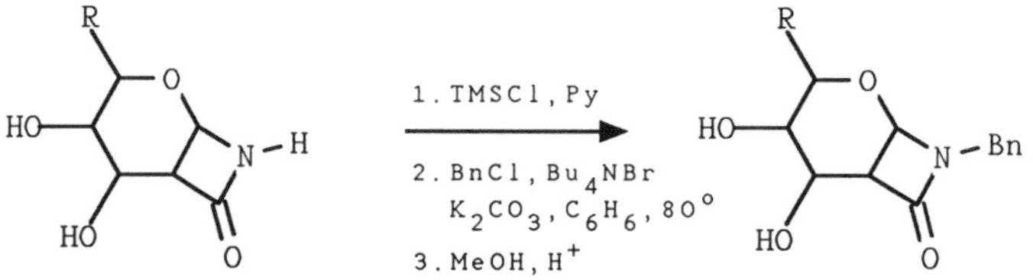

Scheme 15

Glycolic cleavage of the vic-diol grouping present in sugar β-lactams is exemplified in Scheme 17 which shows the oxidation of the β-arabino compound **35**. Periodate oxidation of **35** under standard conditions led to the formation of reactive dialdehyde **36**, which epimerized to the *trans* compound **37**, depending on the pH of the reaction mixture.[24] Upon prolongation of the reaction time, an intramolecular aldol reaction afforded the bicyclic aldehyde **40**. In order to obtain the desired product, control of the reaction time and pH was necessary.[24] The progress of the reaction could be stopped at the stage of 3,4-disubstituted azetidinones by the reduction of the aldehyde functions in **36** or **37** to the respective hydroxymethyl groups (Scheme 17). Alternatively, dialdehyde **36** could be oxidized in a one-pot reaction with sodium chlorite, in the presence of hydrogen peroxide as a chlorine scavanger[25], to afford the dicarboxylic acid **41**.[26] Owing to the malonyl grouping present in **41**, epimerization at the azetidinone occurred readily. At -5 °C, the *cis* configuration was preserved, whereas oxidation at room temperature led only to the *trans* isomer **42** (Scheme 18).[26]

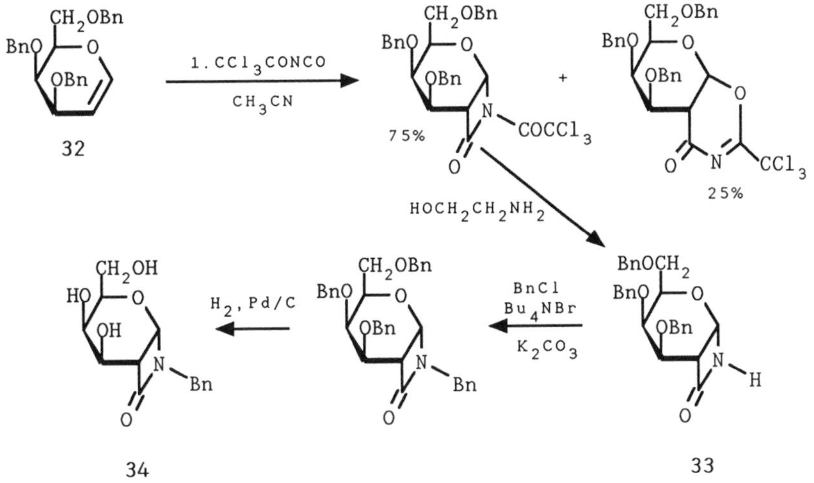

Scheme 16

The usefulness of the above-presented results for the synthesis of 1-oxabicyclic β-lactam antibiotics can be well exemplified by the preparation of the clavam compound **43**.[27] Formation of the clavam skeleton **44** was achieved according to the reaction sequence shown in Scheme 19. This strategy left, however, the unnecessary acetoxymethyl group at the C-6 carbon atom.

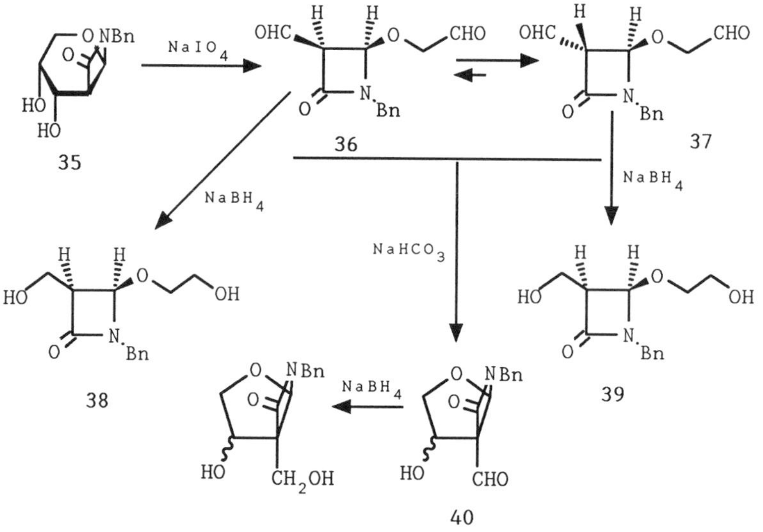

Scheme 17

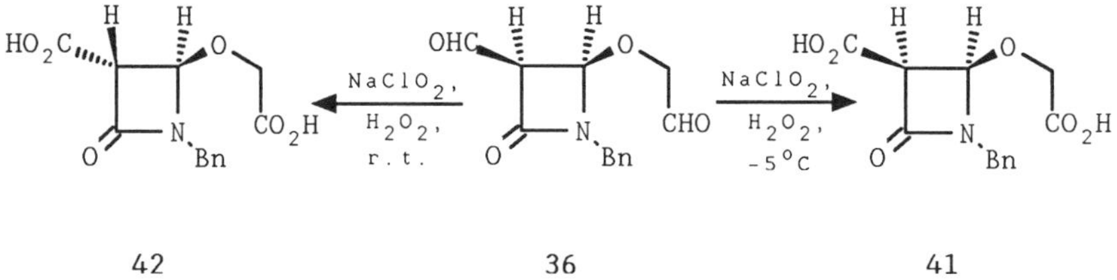

Scheme 18

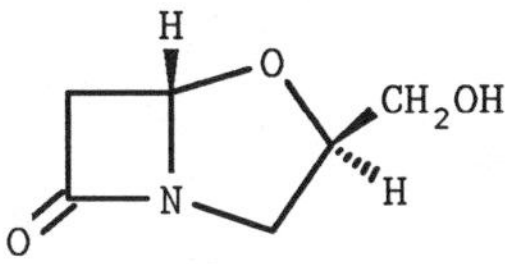

43

We have shown that consideration of the thermodynamic and kinetic properties of the cycloaddition reactions of isocyanates with glycals aided in selection of the most effective substrates and reaction conditions. The glycol cleavage of the vic-diol group present in 2-C:1-N-carbonyl-2-deoxyglycopyranosylamines provides a fully stereocontrolled route to the 1-oxabicyclic β-lactams having desired configuration at the carbon atoms connected to the nitrogen and oxygen atoms. Thus, owing to the stereospecificity of cycloaddition, D-glucal, D-galactal, L-arabinal, and D-xylal gave 3,4-disubstituted azetidinones with the *S*-configuration at the C-4 carbon atom, whereas D-arabinal, L-xylal, L-rhamnal, and L-fucal afforded those of the alternative *R*-configuration.

Construction of a second ring can be achieved in many different ways, one of which, leading to the clavam **44**, is shown (Scheme 19).[28]

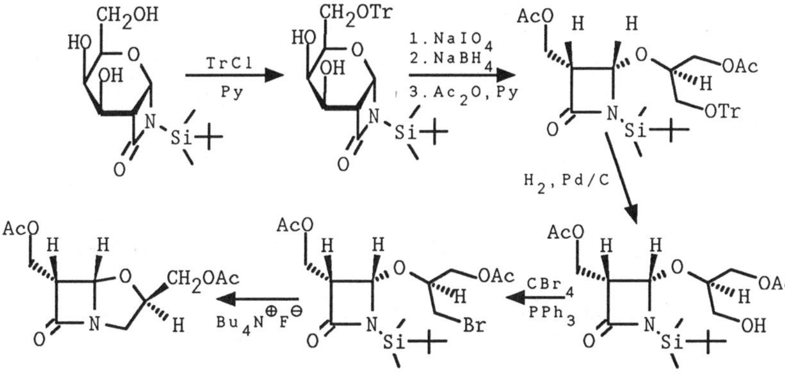

TrCl, Py

1. $NaIO_4$ 2. $NaBH_4$ 3. Ac_2O, Py

H_2, Pd/C

CBr_4, PPh_3

$Bu_4N^{\oplus}F^{\ominus}$

44

Scheme 19

Literature Cited

1. Staudinger, H. *Liebigs Ann. Chem.* **1907**, *356*, 51; Bose, A. K.; Spiegelman, G.; Manhas, M. S. *J. Am. Chem. Soc.* **1968**, *90*, 4506; Wagle, D. R.; Garai, C.; Chiang, J.; Monteleone, M. G.; Kurys, B. E.; Strohmeger, T. W.; Hegde, V. R.; Manhas, M. S.; Bose, A. K. *J. Org. Chem.* **1883**, *53*, 4277; Ghosez, L.; Haveaux, B.; Viehe, H. G. *Angev. Chem., Int. Ed. Engl.* **1969**, *6*, 454; Rogalska, E.; Bełżecki C. *J. Org. Chem.* **1984**, *49*, 1397; Gilman, H.; Spencer, M. *J. Am. Chem. Soc.* **1943**, *65* , 2255; Głuchowski, G.; Cooper, L.; Bergbreiter, D. E.; Newcomb, M. J. *J. Org. Chem.* **1980**, *45*, 3413; Ojima, I.; Inoba, S. *Tetrahedron Lett.* **1980**, *21*, 2077; 2081.
2. Koopel, G. A. in *"The Chemisty of Heterocyclic Compounds. Small Ring Heterocycles"* Ed. Hassner, A.; Wiley, J. & Sons Inc., New York 1983, Vol. *42* Part II, p. 219.
3. Clauss, K.; Grimm, D.; Prossel G. *Liebigs Ann. Chem.* **1974**, 539; Kametani, T.; Honda, T.; Nakayama, A.; Fukumoto, K. *Heterocycles* **1980**, *14*, 1967; Cecchi, R.; Favara, D.; Omodei-Sale, A.; Depaoli, A.; Consonni, P. *Gazz. Chim. It.* **1984**, *114*, 225; Buynak, J. D.; Narayama, Rao M.; Pajouhesh, H.; Yegna Chandrasekaran, R.; Finn, K.; Meester, P.; Chen, S. C. *J. Org. Chem.* **1985**, *50*, 4245.
4. Ohashi, T.; Kan, K.; Sada, I.; Miyama, A.; Watanabe, K. *Eur. Pat. Appl.* **1986**, 167,154; 167,155; Sada, I.; Kan, K.; Ueyama, H.; Matsunobu, S.; Ohashi, T.; Watanabe, K. *Eur. Pat. Appl.* **1988**, 280,962; Ishiguro, M.; Iwata, K.; Nakatsuka, T.; Tanaka, T.; Maeda, Y.; Nishibara, T.; Noguchi, T. *J. Antibiot.* **1988**, *51*, 1685
5. Moriconi, E. J.; Meyer, W. C. *J. Org. Chem.* **1971**, *36*, 2871.
6. Hungerbühler, E.; Bioloz, M.; Ernest, I.; Kalvoda, J.; Langa, M.; Schneider, P.; Sedelmeier, G. in Yoshida, Z.; Shiba, T.; Ohshiro Y. Ed. *"New Aspects of Organic Chemistry"* VCh, New York, 1989, p. 419.
7. Graf, R. *Liebigs Ann. Chem.* **1963** *661*, 111; Moriconi, E. J.; Crawford, W. C. *J. Org. Chem.* **1968**, *33*, 370; Graf, R. *Org. Synth.* **1966** *46*, 51.
8. Hall, R. H.; Jordaan, A.; Lourens, G. J. *J. Chem. Soc., Perkin I* **1973**, 38; Hall, R. H.; Jordaan, A.; de Villiers, O. G. ibid. **1975**, 626.
9. Effenberger, F.; Gleiter, R. *Chem. Ber.* **1964**, *97*, 1576.
10. Barrett, A. G. M.; Fenwick, A. and Betts, M. J. *J. Chem. Soc., Chem. Commun.* **1983**, 299.
11. Chan, J. H.; Hall, S. S. *J. Org. Chem.* **1984**, *49*, 195.

12. Chitwood, J. L.; Gott, P. G.; Martin, J. C. ibid. **1971**, *36*, 2228.
13. Isaacs, N. S.; Rannola, E. *J. Chem. Soc., Perkin II*, **1975**, 1555; Swieton, G.; von Jouanne, J.; Keim, H.; Huisgen, R. ibid., **1983**, 37; Wiering, P. G.; Steinberg, H. *Recl. J. Roy. Neth. Chem. Soc.* **1981**, *100*, 13.
14. Jenner, G. in *"Organic High Pressure Chemistry"*; Ed. Le Noble, W. J. Elsevier, Amsterdam 1988, p. 143, and references cited therein.
15 Chmielewski, M.; Kałuża, Z.; Bełżecki, C.; Sałański, P.; Jurczak, J. *Tetrahedron Lett.* **1984**, *25*, 4797.
16. Chmielewski, M.; Kałuża, Z.; Bełżecki, C.; Sałański, P.; Jurczak, J.; Adamowicz, H. *Tetrahedron* **1985**, *41*, 2441.
17. Chmielewski, M.; Kałuża, Z. unpublished results.
18. Chmielewski, M.; Kałuża, Z.; Mostowicz, D.; Bełżecki, C.; Baranowska, E.; Jacobsen, J.P.; Sałański, P.; Jurczak, J. *Tetrahedron* **1987**, *43*, 4563.
19. Chmielewski, M.; Kałuża, Z. *J. Org. Chem.* **1986**, *51*, 2395.
20. Chmielewski, M.; Kałuża, Z. *Carbohydr. Res.* **1987**, *167*, 143.
21. Chmielewski, M.; Kałuża, Z.; Abramski, W.; Mostowicz, D.; Hintze, B.; Bełżecki C. *Bull. Acad. Pol. Sci.* **1987**, *35*, 245.
22. Bełżecki, C.; Grodner, J.; Urbański, R.; Chmielewski, M. unpublished results.
23. Chmielewski, M.; Badowska-Rosłonek, K. unpublished results.
24. Chmielewski, M.; Kałuża, Z.; Abramski, W.; Grodner, J.; Bełżecki, C.; Sedmera, P. *Tetrahedron* **1989**, *45*, 227.
25. Dalcanale E.; Montanari F. *J. Org. Chem.* **1986**, *51*, 567.
26. Chmielewski, M.; Grodner, J.; Fudong Wang unpublished results.
27. Brown, D.; Evans, J. R.; Fletton, R. A. *J. Chem. Soc., Chem. Commun.* **1979**, 282.
28. Chmielewski, M.; Grodner, J. unpublished results.

RECEIVED December 2, 1991

Chapter 5

Diels–Alder Cycloaddition to Unsaturated Sugars

Stereocontrol as a Function of Structure and Stereochemistry

Derek Horton, Dongsoo Koh, Yasushi Takagi, and Takayuki Usui

Department of Chemistry, Ohio State University, Columbus, OH 43210

Direct Wittig reaction of Ph_3PCHCO_2Me with the four unsubstituted D-aldopentoses followed by acetylation provides convenient preparative access to acyclic seven-carbon trans-2,3-unsaturated sugar derivatives. These products served as dienophiles for a detailed comparative study in Diels—Alder cycloaddition with cyclopentadiene. Related syntheses afforded analogous cis-dienophiles. Cycloaddition under uncatalyzed thermal conditions gave mixtures of the four possible stereoisomeric norbornene adducts. The endo, exo ratios, and diastereofacial selectivities of the adducts were determined by NMR spectroscopy and by chemical transformations, supplemented by selected X-ray crystallographic analyses. Different distributions of isomers were encountered when a Lewis acid was used to catalyze the cycloaddition. The reaction can be controlled to provide preparative access to selected isomers and thus constitutes a versatile method for chirality transfer from the precursor sugar to four new asymmetric centers in a carbocyclic framework.

As part of a general program (1—4) on synthetic transformations of sugars having potential value for access to enantiomerically pure, polysubstituted carbocycles, we have made a systematic study of the reactions of various dienes with sugar-derived alkenes serving as dienophiles. Here we present the results of reactions between cyclopentadiene and a complete stereoisomeric set of seven-carbon trans-2,3-unsaturated aldonic esters derived from the D-aldopentoses.

Comparative results with corresponding seven-carbon cis-dienophiles are also included. The work allows

0097–6156/92/0494–0066$06.00/0

predictive understanding of the steric factors dictating the product distribution in the reaction. Further, it provides a potential methodology for using readily available sugars as chiral precursors for obtaining tetra-C-substituted cyclopentane derivatives, with each substituent capable of differential chemical elaboration, enantiomerically pure in all sixteen possible stereoisomeric forms.

Synthesis of the trans Dienophiles

The four D-aldopentoses (D-ribose, **1**, D-arabinose, **2**, D-lyxose, **3**, and D-xylose, **4**) were converted into the corresponding methyl (E)-4,5,6,7-tetra-O-acetyl-2,3-dideoxyhept-2-enonates (**5**—**8**).

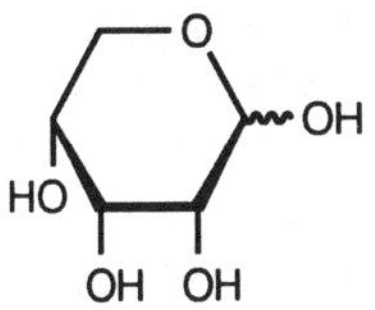

HO OH OH OH O **1**

HO HO OH OH O **2**

HO OH HO OH O **3**

HO OH OH OH O **4**

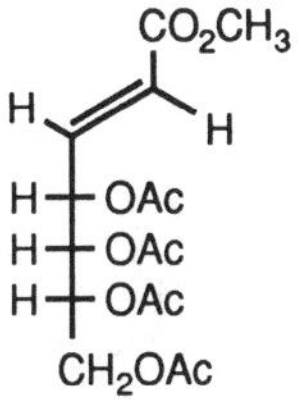

CO_2CH_3, H, H, H—OAc, H—OAc, H—OAc, CH_2OAc **5**

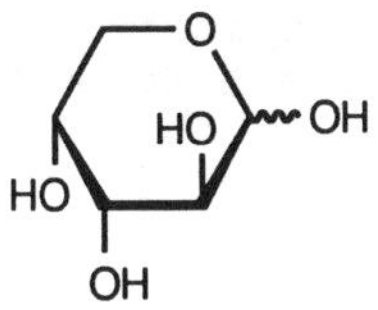

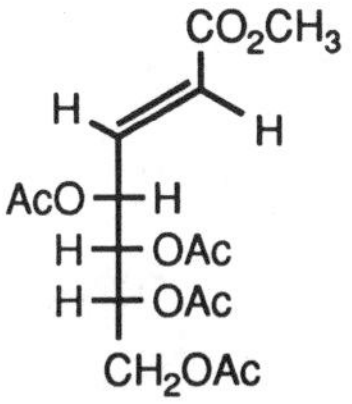

CO_2CH_3, H, H, AcO—H, H—OAc, H—OAc, CH_2OAc **6**

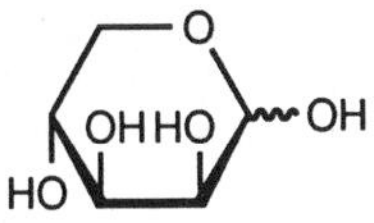

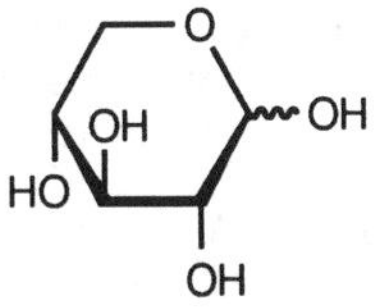

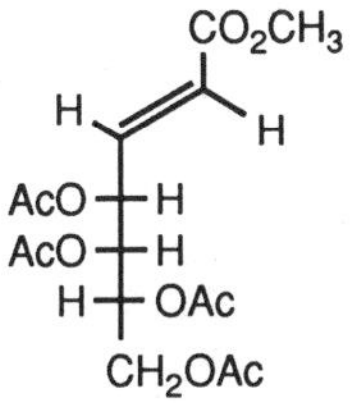

CO_2CH_3, H, H, AcO—H, AcO—H, H—OAc, CH_2OAc **7**

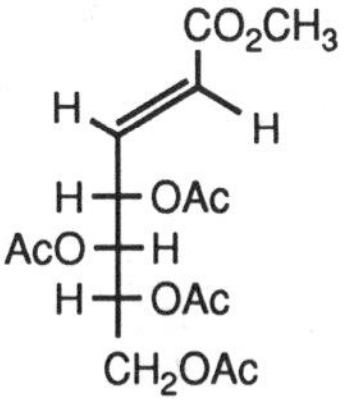

CO_2CH_3, H, H, H—OAc, AcO—H, H—OAc, CH_2OAc **8**

The pentoses may be converted by the conventional sequence (5) of dithioacetal formation, acetylation, and demercaptalation into the respective aldehydo-pentose 2,3,4,5-tetraacetates, which afford by Wittig chain-extension (2) the desired dienophiles. Thus aldehydo-D-arabinose 2,3,4,5-tetraacetate reacted with Ph_3PCHCO_2Me in boiling benzene to give 90% of the pure, crystalline E-unsaturated sugar derivative **6**, and the corresponding enantiomer was likewise prepared from L-arabinose. However, the preparation required several steps and the net yield from arabinose was only 23%.

A significant preparative improvement was realized by conducting the Wittig reaction directly (6) on the free aldopentose (**2**) with the ylid Ph_3PCHCO_2Me in boiling THF. The initial heterogeneous mixture became a clear solution. Acetylation of the resultant mixture of products and flash chromatography afforded the pure alkene **6** in 26% yield as the fastest-moving product. This yield could be boosted to 50% by incorporation of $Cu(OAc)_2$ in the initial reaction medium. Other products formed in the reaction result from internal Michael addition from the C_7 alkene prior to the acetylation step, and their formation is significantly attenuated when the copper salt is incorporated in the reaction medium.

The same direct reaction of the aldopentose with Ph_3PCHCO_2Me allowed conversion of D-ribose (**1**) into **5** in 25% yield; here the use of $Cu(OAc)_2$ suppressed rather than enhanced conversion into the alkene. The direct Wittig reaction with D-lyxose (**3**) was the most successful among the four pentoses and afforded the alkene **7** in 61% yield as the major product; its yield was depressed to 40% when $Cu(OAc)_2$ was present. With D-xylose (**4**) the direct reaction gave the alkene **8**; the yield of 43% was increased to 49% when $Cu(OAc)_2$ was present in the reaction mixture.

Related trans dienophiles were prepared by the same direct Wittig reaction on the free aldopentoses but with an acetonation step in place of acetylation; thus prepared were the 4,5:6,7-di-O-isopropylidene analogue (**9**) of **5** (yield 29%), the analogue **10** of **6** (20%), the analogue **11** of **7** (34%), and the analogue **12** of **8** (18%).

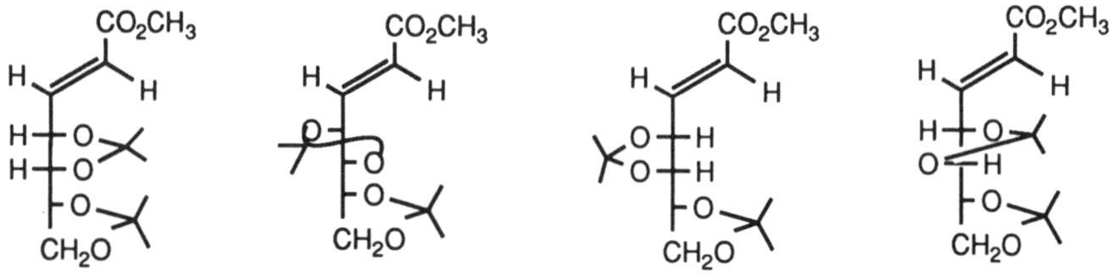

9 **10** **11** **12**

Additional trans dienophiles were prepared by the direct Wittig reaction from free and from partially protected aldopentoses. 2-Deoxy-D-erythro-pentose (**13**) gave 73% of an 11:1 separable mixture of the E and Z C_7 alkenes (**14** and **15**, respectively), with no accompanying cyclized products.

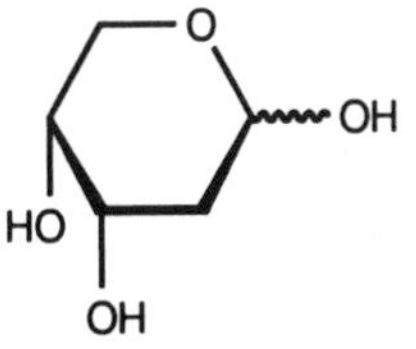

13 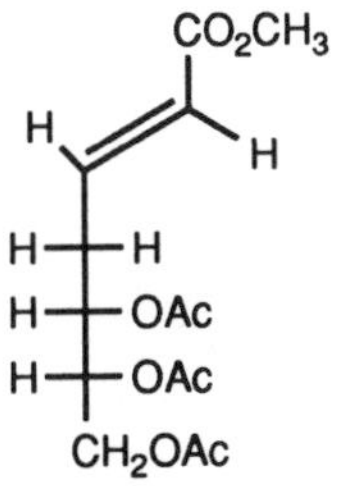 14 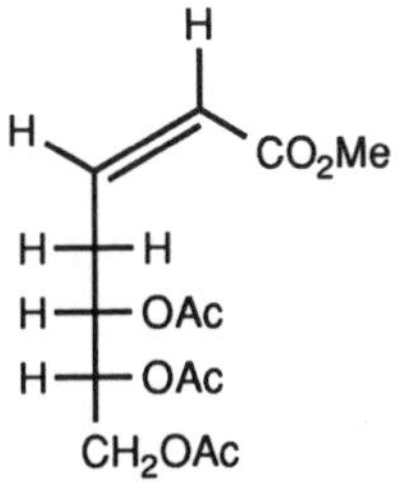 15

Kinetic acetonation (7) of D-ribose (**1**) to give 3,4-O-isopropylidene-D-ribopyranose (**16**), followed by the direct Wittig reaction gave mainly (85%) the E product **17**. In contrast, the thermodynamic acetonation product, 2,3-O-isopropylidene-D-ribofuranose (**18**) afforded mainly (~70%) the Z (cis) ester, isolated as its 6,7-diacetate **19**.

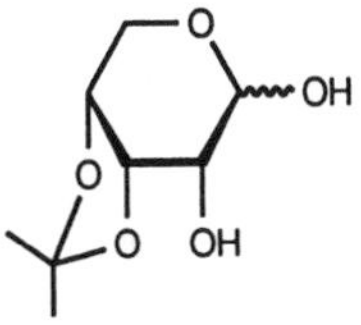

16 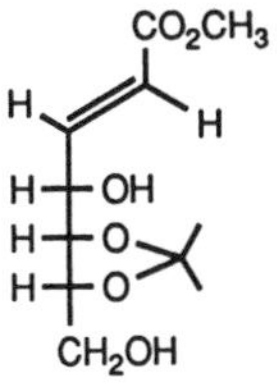 17 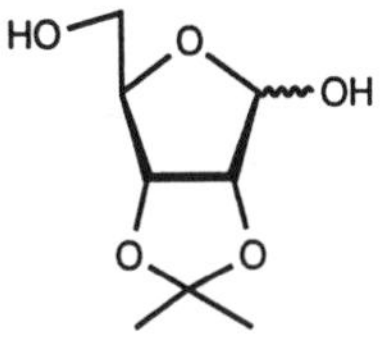 18 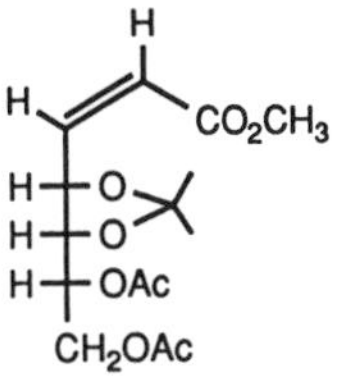 19

Synthesis of the cis Dienophiles

In addition to compounds **15** and **19**, other dienophiles having the cis configuration were obtained from aldehydo-D-arabinose 2,3,4,5-tetraacetate by Horner—Emmons alkenation with $(CF_3CH_2O)_2POCH_2CO_2Me$, which afforded the crystalline Z alkene **20** in 91% yield. The reaction of 2,3;4,5-di-O-isopropylidene-aldehydo-D-arabinose with Ph_3PCHCO_2Me in methanol at room temperature gave a 1:4 E,Z mixture of alkenes which, on treatment with TsOH in methanol followed by acetylation afforded in 39% net yield the crystalline butenolide **21**.

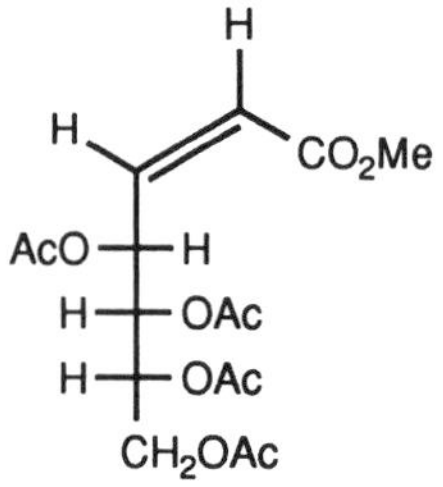

20

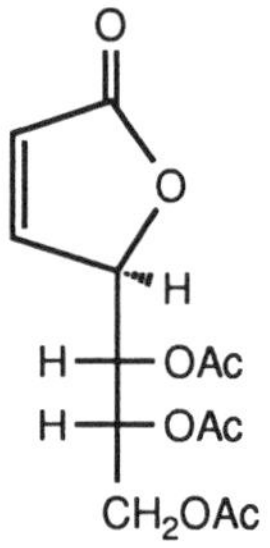

21

Reaction of the trans Dienophiles with Cyclopentadiene

General. All of the alkenes prepared as dienophiles reacted with cyclopentadiene under thermal conditions (2,3) to afford in high net yield a mixture of Diels—Alder adducts. Four stereoisomers (**A**—**D**) are, in principle, possible in each instance through transition states leading to endo or exo carboxylate norbornene products and from si or re face attack by the diene on the dienophile. The designation si or re refers to the Cahn—Ingold—Prelog priorities at the double bond (lowest-numbered asymmetric position in the starting sugar derivative) of the dienophile.

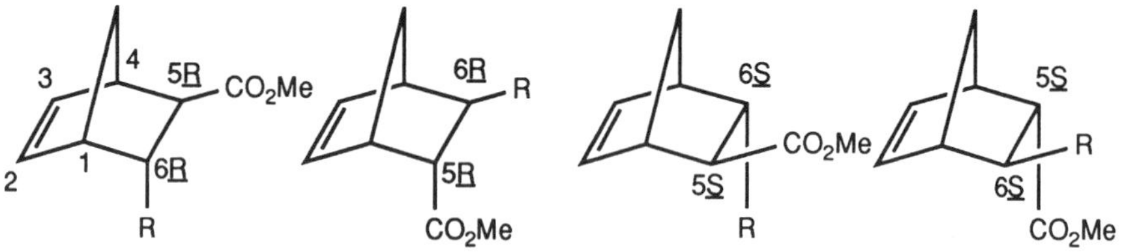

A (exo) **B** (endo) **C** (exo) **D** (endo)

The observed products from the trans dienophiles generally comprised mixtures of all four possible isomers, which could be detected in most instances by differences in TLC mobility, and be quantitated by NMR integration of key signals (the proton α to the carboxylate group and the methyl resonance of that group). In some cases direct separation of all four isomers was perfomed by careful chromatography. A general feature in all of the reactions was that the exo-carboxylate product is favored, and the facial selectivity is that corresponding to orientation of the allylic oxygen atom of the dienophile toward the diene in the transition state of a concerted process (see following scheme). The products are the result of a kinetically controlled process, as evidenced by the fact that single isolated isomers from the reactions, when subjected to the same conditions used for the Diels—Alder reaction, were recovered unchanged.

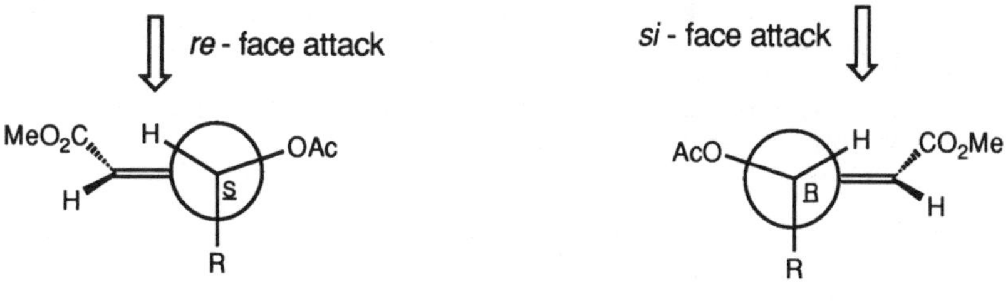

re - face attack

si - face attack

5 (D-*ribo*)
8 (D-*xylo*)

6 (D-*arabino*)
7 (D-*lyxo*)

Diastereofacial Selectivities of Acetylated Dienophiles

Preparative Utility. Because of the ready availability of both enantiomers of arabinose, the reactions of the derived C_7 dienophiles were evaluated in particular detail for potential preparative utility, with emphasis on the ability to separate by crystallization the major isomer from the reaction. Thus dienophile **6** reacted to give, by direct isolation, a 40% yield of pure, crystalline methyl (5S,6S)-6-*endo*-(1,2,3,4-tetra-*O*-acetyl-D-*arabino*-tetritol-1-yl)bicyclo[2.2.1]hept-2-eno-5-*exo*-carboxylate (**22**), mp 103.5—104.5°, $[\alpha]_D$ +73°, and likewise the L enantiomer of **6** gave methyl (5R,6R)-6-*endo*-(1,2,3,4-tetra-*O*-acetyl-L-*arabino*-tetritol-1-yl)bicyclo[2.2.1]hept-2-eno-5-*exo*-carboxylate (**23**).

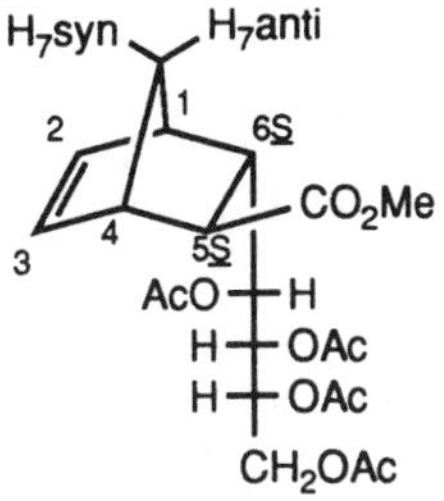

22

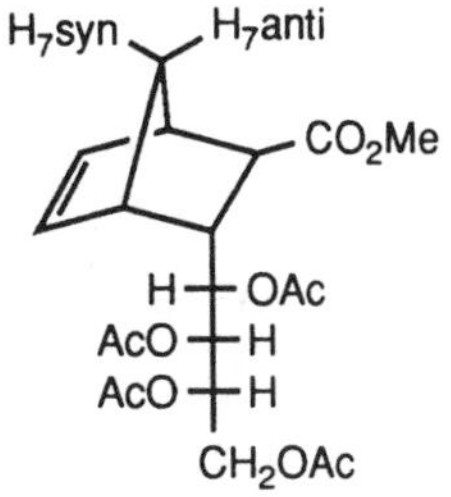

23

Proof of Structure. Structural and stereochemical identification of the products was achieved by NMR, by chemical transformations, and was further substantiated by X-ray crystallography of selected products. Thus the norbornene derivative **22** was identified as an *exo* carboxylate derivative by the relatively high-field location of the *endo* disposed H-5 signal (proton α to the carboxylate group). Sequential *O*-deacetylation to the tetrol **24**, periodate degradation of the polyhydroxy chain to the aldehydo ester **25**, and reduction with $LiAlH_4$ afforded a product (**26**) identified as (5*S*,6*S*)-2-norbornene-5,6-dimethanol (*2,8*), $[\alpha]_D$ -23°, thus establishing the chirality of **22** as 5*S*,6*S*.

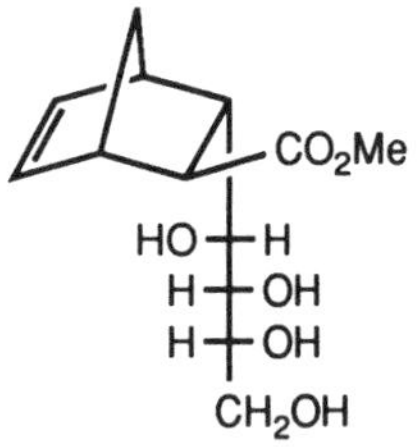

24 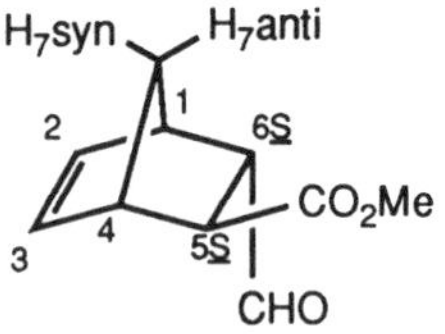 **25** 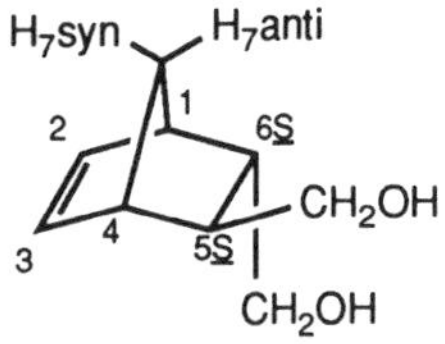 **26**

The X-ray crystal structure (*1*) of **22**, taken in conjunction with the known D-*arabino* stereochemistry of the starting sugar, provided independent affirmation of the absolute stereochemistry of the product (and thus also of its enantiomer **23**) as well as assurance that no stereochemical alteration of the chiral centers in the original sugar chain had taken place, and provided a firm point of reference for assigning structures to all of the remaining isomers formed in the cycloaddition reaction. By extension, the NMR parameters for the four isomers from this reaction established a basis for structural attribution of the products from cycloaddition of cyclopentadiene to the other dienophiles in this study.

Lewis Acid-Catalyzed Cycloaddition. When the cycloaddition to the D-arabinose-derived dienophile **6** was performed in the presence of a Lewis acid catalyst ($AlCl_3$), both the geometric and the facial selectivities were reversed and the major product, isolated crystalline in 36% yield was methyl (5*R*,6*R*)-5-*exo*-(1,2,3,4-tetra-*O*-acetyl-D-*arabino*-tetritol-1-yl)bicyclo[2.2.1]hept-2-eno-6-*endo*-carboxylate (**27**), the *endo*-carboxylate product resulting from attack on the face of the diene *opposite* the allylic acetoxy group. Similar reaction of the L-arabinose-derived dienophile correspondingly gave methyl (5*S*,6*S*)-5-*exo*-(1,2,3,4-tetra-*O*-acetyl-*L*-*arabino*-tetritol-1-yl)bicyclo[2.2.1]hept-2-eno-6-*endo*-carboxylate (**28**).

In consequence, thus, by using both enantiomers of arabinose and reaction under either thermal or Lewis acid-catalyzed conditions, it is possible to prepare the four possible 5,6-trans norbornene products in enantiomerically pure form. These can be used, in turn, to prepare tetra-C-substituted cyclopentane derivatives in which all of the four carbon substituents are capable of differential synthetic elaboration. Details of such transformations are documented in a separate report (4).

Quantitative Distribution of Adducts as a Function of Dienophile Stereochemistry. For accurate quantitation of the isomer distribution in the products of cycloaddition to each of the dienophiles **5—8**, the entire mixtures of the four stereoisomeric products in each instance were first subjected to sequential O-deacetylation and periodate oxidation to afford a mixture of two aldehydo esters **29** and **30**, which upon reduction with $LiAlH_4$ afforded trans-2-norbornene-5,6-dimethanol as an unequal mixture of the two enantiomers (only the 5S,6S enantiomer is shown). NMR analysis of the mixture of **29** and **30** showed distinctive resonances for the CH_3O and CHO groups in exo and endo orientations, permitting accurate determination of the endo/exo ratio of the products in the mixture. The observed specific rotation of the diol, in comparison with that (+23°) determined for the enantiomerically pure 5S,6S diol **26** (and its enantiomer), provided a quantitative measure of the si,re diastereofacial selectivity.

The results for the four dienophiles **5—8**, together with that for the 2-deoxy-D-erythro-pentose-derived dienophile **14**, are shown in Table I.

Table I. Stereoselectivity in the reaction of acetylated dienophiles with cyclopentadiene

Dienophile	Configuration	Yield (%)	endo/exo	si/re
5	D-ribo	86	31:69	30:70
6	D-arabino	92	31:69	64:36
7	D-lyxo	97	37:63	60:40
8	D-xylo	72	29:71	38:62
14	(4-deoxy)-D-erythro	93	41:59	53:47

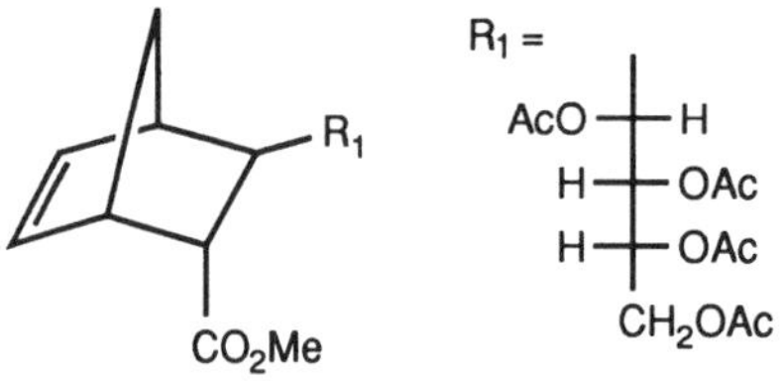

R_1 =
AcO–H
H–OAc
H–OAc
CH_2OAc
CO_2Me
27

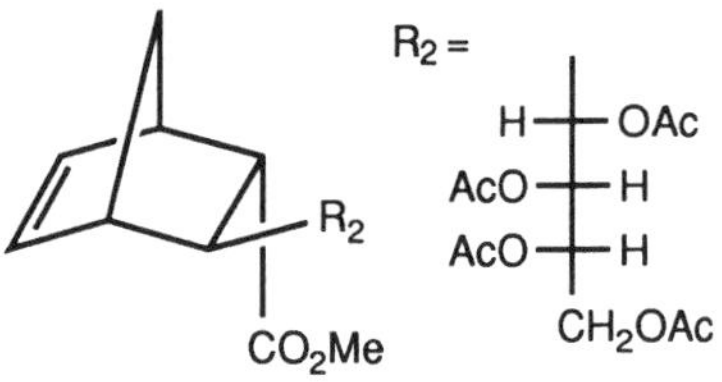

R_2 =
H–OAc
AcO–H
AcO–H
CH_2OAc
CO_2Me
28

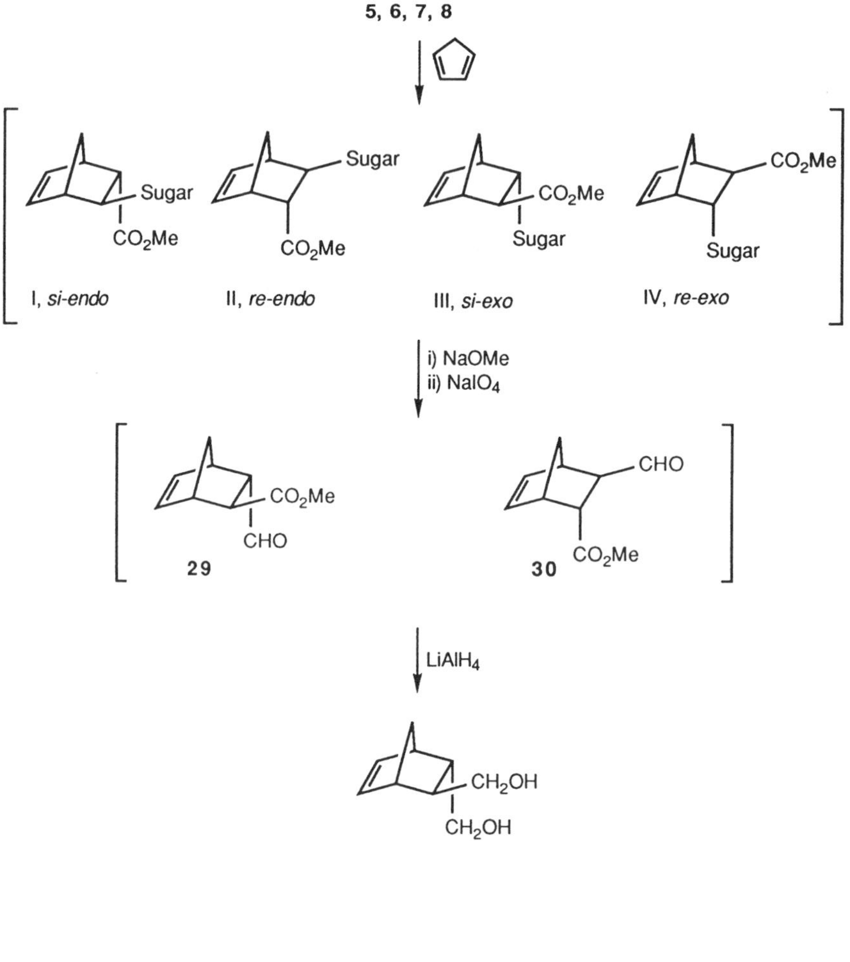

5, 6, 7, 8

Sugar
CO_2Me
I, *si-endo*

Sugar
CO_2Me
II, *re-endo*

CO_2Me
Sugar
III, *si-exo*

CO_2Me
Sugar
IV, *re-exo*

i) NaOMe
ii) $NaIO_4$

CO_2Me
CHO
29

CHO
CO_2Me
30

$LiAlH_4$

CH_2OH
CH_2OH

It is evident from these results that the thermal reaction favors the exo carboxylate products throughout. As regards diastereofacial selectivity, two of the dienophiles (**5** and **8**), which have the same S configuration at the allylic center, show the same tendency for favored attack at the re face. In contrast, the other two dienophiles (**6** and **7**, R configuration at the allylic center) show, as expected, favored si-face attack. The dienophile **14**, having no chiral group at the allylic position, showed negligible facial selectivity.

The relatively low diastereofacial selectivities exhibited by the acyclic sugar-chain enonates **5**—**8** may be ascribed to the conformational mobility of the chain. These chains are depicted in their Fischer projections rather than as conformational representations for the specific reason that the planar zigzag orientation of the chain is clearly favored only in chains having the arabino stereochemistry (9). For the other configurations, the conformational preference is for non-extended conformations that may be conformational mixtures separated by low energy-barriers. It is clearly naive to depict exact molecular orientations for putative transition states in such reactions. Nevertheless, the model depicted here for interpreting the course of the reaction, which is in accord with the general model proposed by Trost (10) for diastereofacial selectivity in additions to alkenes having an adjacent asymmetric center, has predictive utility in these reactions.

The isopropylidenated enonates **9**—**12** have less conformational freedom than the tetraacetates **5**—**8** and were thus expected to show higher diastereofacial selectivities in the Diels--Alder reaction. This was borne out in the experimental data. The D-lyxo enonate **11** in particular gave a readily separable mixture after reaction with cyclopentadiene under thermal conditions, and afforded a 55% isolated yield of the crystalline si-exo product, mp 88—89°, along with 30% of the si-endo product. The structures of these products were confirmed as before by degradative sequences. The diastereofacial selectivity was >9:1 in favor of si-face attack.

For the other isopropylidenated dienophiles **9**, **10**, and **12**, separation of the product isomers was more difficult, but determination of the ratios of the four possible products as before gave the results shown in Table II.

Table II. Stereoselectivity in the Reaction of Isopropylidenated Dienophiles with Cyclopentadiene

Dienophile	Configuration	Yield (%)	endo/exo	si/re
9	D-ribo	86	31:69	18:82
10	D-arabino	98	35:65	62:38
11	D-lyxo	97	40:60	86:14
12	D-xylo	96	43:57	32:68
Methyl (E)-2,3-dideoxy-4,5-O-isopropylidene-D-glycero-pent-2-enonate (11)		87	40:60	32:68

The preference for formation of exo products is again evident throughout. As for diastereofacial selectivty, the pattern shown with the acetylated dienophiles **5**—**8** was again evident, with favored re-face attack for those compounds (**9** and **12**) having the R configuration at the allylic center and si-face attack for those (**10** and **11**) having the S configuration at the allylic center. However, with these isopropylidenated dienophiles, the diastereofacial selectivity was much higher than with the acetylated analogues, especially in the ribo and lyxo isomers, a factor of importance in any proposed application in chiral synthesis.

These results again accord with predictions based on the Trost model (10), and the following schematic illustration satifactorily interprets the observed behavior of compounds **9**—**12** as well as that of the 5-carbon lower homologue also listed in Table II and the subject of a recent independent study (11).

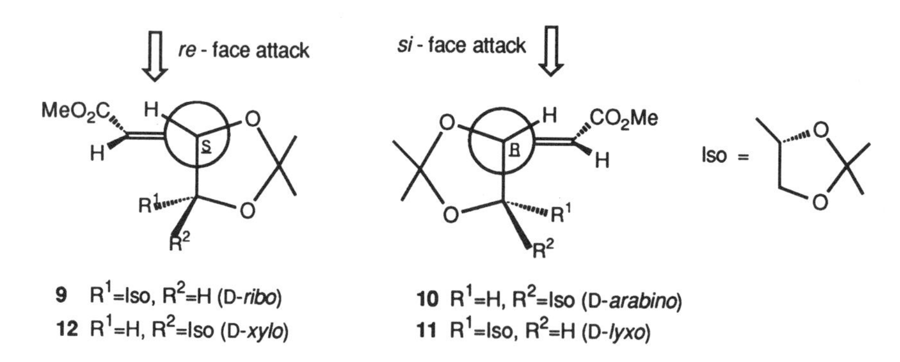

Diastereofacial Selectivities of Acetonated Dienophiles

Reaction of the cis Dienophiles with Cyclopentadiene

A very high degree of asymmetric induction was evident in the thermal reaction of the D-arabinose-derived cis-dienophile **20** with cyclopentadiene, and the crystalline adduct isolated in 95% yield was identified as methyl (5R,6S)-6-endo-(1,2,3,4-tetra-O-acetyl-D-arabino-tetritol-1-yl)bicyclo[2.2.1]hept-2-eno-5-endo-carboxylate (**31**), mp 103° by NMR spectroscopy and also by X-ray crystallography. Use of the same sequence starting from L-arabinose gave the enantiomer (**32**) of **31**.

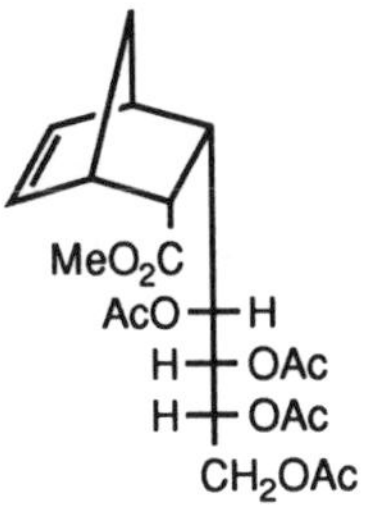

31 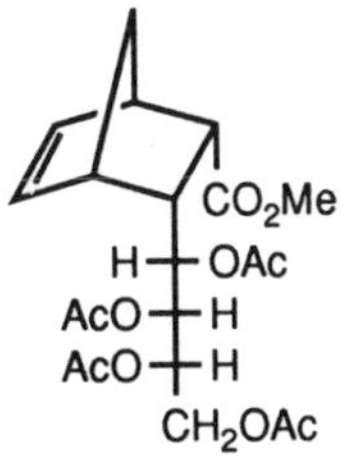 **32**

A very small proportion of a second isomer isolated from the reaction with **20** was identified as the exo 5R,6S isomer of **31**.

The observed high diastereofacial selectivity is attributable to conformational restriction at the allylic center in dienophile **20**. The steric bulk of the sugar chain and the acetyl group effectively limit the C-3—C-4 rotamers to a single conformer, as shown in the following scheme. The diene attacks almost exclusively from the same side as the allylic oxygen atom (si-face for the D-enonate **20**), where the steric hindrance is lowest.

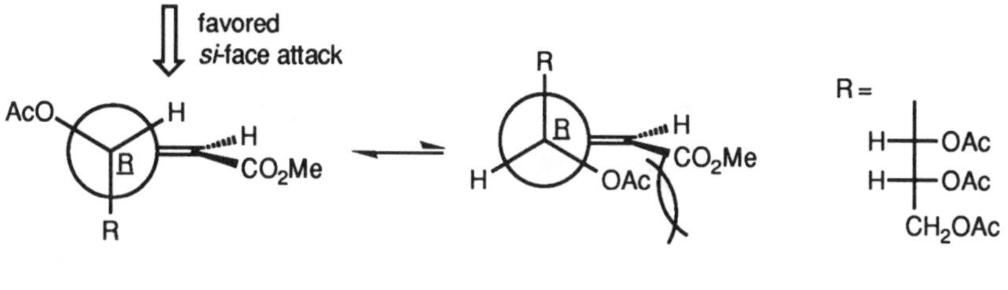

20A favored

20B disfavored

The butenolide **21** provides an opportunity to test the foregoing hypothesis in that the lactone ring locks the dienophile in the conformation that is disfavored for the acyclic dienophile **20**. The results of the cycloaddition reaction with cyclopentadiene are entirely supportive in that the observed products are exclusively those of re-face attack. The reaction gave two products only. The major one was (5S,6R)-6-endo-(2,3,4-tri-O-acetyl-D-arabino-tetritol-1-yl)bicyclo[2.2.1]hept-2-ene-5-endo-carboxylic 1,4-lactone (**33**), isolated crystalline in 70% yield, mp 149°. The minor product, obtained in 11% yield, was the exo (5S,6R) isomer **34**.

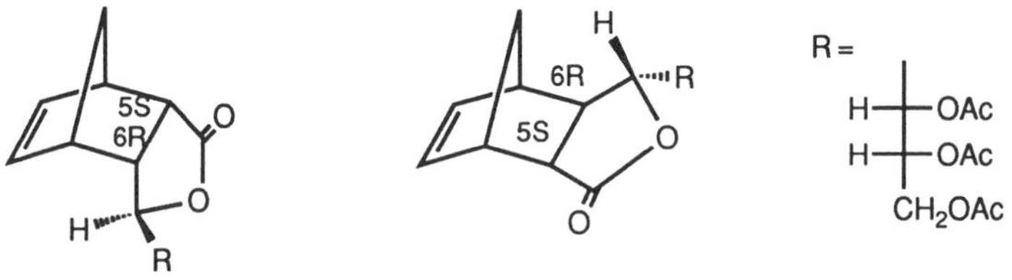

33 (70%) **34** (11%)

Structures of the products were again established by NMR spectroscopy, and that of **33** firmly consolidated by X-ray crystallography. The lactone ring in butenolide **21** constrains the groups adjacent to the double bond into the arrangement depicted as the unfavorable conformation in the acyclic cis dienophile **20**, and consequent approach by cyclopentadiene from the exposed face results in exclusive re-face attack. The consistency in the diastereofacial selectivities for **20** and **21** supports the hypothesis of conformational restriction in the acyclic dienophile **20** and its analogues.

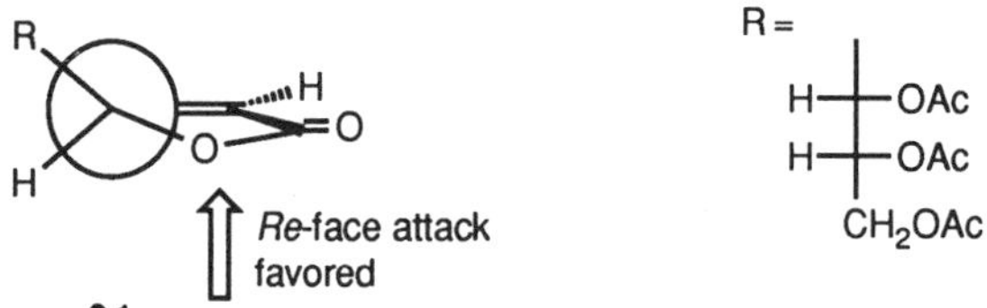

21

The cis-enonate **15**, which is deoxygenated at the allylic position, reacted with cyclopentadiene to give a mixture of all four possible adducts. The diastereofacial selectivity was negligible and the endo,exo ratio (6.5:1) was similar to that observed with butenolide **21**. This result demonstrates that the stereocenter at the allylic position alone determines the diastereofacial selectivity of the Diels—Alder reaction with these acyclic cis dienophiles.

Although this report has focused principally on dienophiles not constrained by a ring system, one final example is the addition of cyclopentadiene to a 6-membered dieneophile, 4-O-acetyl-2,3,6-trideoxy-L-erythro-hex-2-enonate 1,5-lactone (**35**, 12). Reaction of **35** with cyclopentadiene gave principally the two endo products **36** and **37** in 31 and 38% yields, respectively, with only 12% of a mixture of the two exo products being formed.

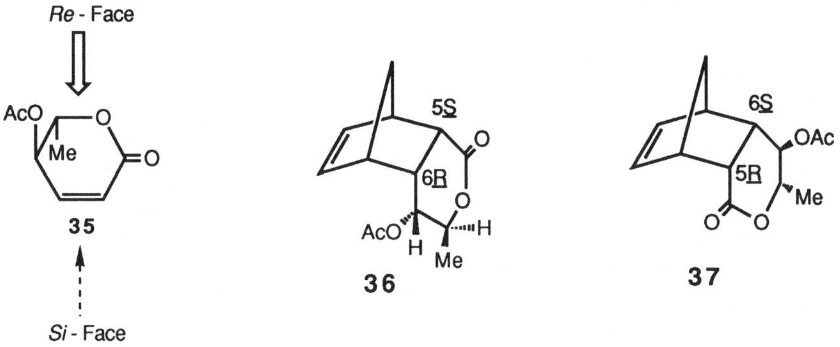

Re - Face

AcO O Me O

35

Si - Face

5S O 6R O AcO H H Me

36

6S OAc 5R Me O O

37

The very low diastereofacial selectivity observed with lactone **36** may be ascribed to the competing effects of two chiral centers, the allylic oxygen atom O-4 (favored si-face attack) and the C-5 methyl group (favored re-face attack).

Summary and Conclusions

These studies of asymmetric Diels—Alder reactions with α,ß-unsaturated sugar enonates permit the following generalizations.

The diastereofacial selectivities are mainly controlled by the allylic configuration.

Compounds having the same configuration at the allylic position undergo attack on the dienophile at the same face.

The cis dienophiles generally exhibit greater diastereofacial selectivity than the trans analogues.

The diastereofacial selectivities observed are satisfactorily rationalized by Trost's steric model.

The reaction has broad potential in synthesis of enantiomerically pure, polysubstituted carbocycles by chirality transfer from sugar precursors.

Literature Cited

1. Horton, D.; Machinami, T.; Takagi, Y.; Bergmann, C. W.; Christoph, G. D., J. Chem. Soc., Chem. Commun., 1983, 1164—1166.
2. Horton, D.; Machinami, T.; Takagi, Y., Carbohydr. Res., 1983, 121, 135—161.
3. Horton D,; Usui, T., Carbohydr. Res., 1991, 216, 33—49.
4. Horton, D.; Usui, T., Carbohydr. Res., 1991, 216, 51—59.
5. Wolfrom, M. L.; Weisblat, D. I.; Zophy, W. H.; Waisbrot, S. W., J. Am. Chem. Soc., 1941, 63, 201—203.
6. For early work on Wittig extension of sugar chains, see Zhdanov, Yu A.; Alexeev, Yu. E.; Alexeeva, V. G., Adv. Carbohydr. Chem. Biochem., 1972, 27, 227—299.
7. Gelas, J.; Horton, D., Heterocycles, 1981, 16, 1587—1601.
8. Takano, S.; Kurotaki, A., Synthesis, 1987, 1075—1078.
9. Blanc-Muesser, M.; Defaye, J.; Horton, D., Carbohydr. Res., 1980, 87, 71—86 and earlier papers cited therein.
10. Trost, B.M.; Lynch, J.; Renaut, P., Tetrahedron Lett., 1985, 26, 6313—6316; compare Tripathy, R.; Franck, R. W.; Onan, K.D., J. Am. Chem. Soc., 1988, 110, 3257—3262.
11. Krief, A.; Dumont, W.; Pasau, P,; Lecomte, P., Tetrahedron, 1989, 45, 3039—3052.
12. Jarglig, P.; Lichtenthaler, F. W., Tetrahedron Lett., 1982, 3781—3784.

RECEIVED February 4, 1992

Chapter 6

The [4 + 2] Cycloaddition of Azodicarboxylates and Cyclic Vinyl Ethers

Application to the Synthesis of Simple and Complex 2-Aminosaccharides

Yves Leblanc and Marc Labelle

Merck Frosst Centre for Therapeutic Research, P.O. Box 1005, Pointe Claire-Dorval, Quebec H9H 3L1, Canada

The [4+2] cycloaddition reaction of furanoid and pyranoid glycals with azodicarboxylates allows the introduction of a nitrogen atom at C-2 of a carbohydrate, and the adducts thus obtained can be converted to structurally complex 2-amino saccharides under mild conditions. Towards this end, dibenzyl and bis(2,2,2-trichloroethyl) azodicarboxylates were used as dienes since they allow the free amines to be generated in a single operation under mild conditions later in the synthetic sequence.

Amino sugars are broadly distributed in nature. They are fundamental constituents of several biologically important molecules such as antigenic determinants *(1)*, glycoproteins *(2)*, glycolipids, and antibiotics such as tunicamycin (*3*) and gentamycin (*4*). Amino sugars are synthetically challenging molecules and only a few methods exist for their preparation. These methods are generally specific to certain substrates and in some cases lead to mixtures of regio and/or stereoisomers (*5-8*). Recently, new synthetic methods (*9*) have appeared in which both the introduction of the amino group at C-2 and the formation of the glycosidic linkage can be achieved in a stereoselective fashion. In this chapter, we will present our approach to the preparation of 2-amino 2-deoxy carbohydrates (*10-12*).

Our approach is based on a reaction independently reported by two research groups (*13-20*). It has been found that azodicarboxylates add to vinyl ethers to provide [4+2] and/or [2+2] cycloadducts. They have shown that the [4+2] cycloaddition reaction is accelerated upon heating or irradiation at 350 nm, presumably due to a partial isomerization of the trans azo compound to the cis isomer, which would be more reactive. In principle, this reaction should be equally efficient with glycals, thereby allowing the introduction of an amino

0097–6156/92/0494–0081$06.00/0

function at C-2 of a carbohydrate backbone (Figure 1). Furthermore, the rigidity of glycals with a properly oriented C-3 substituent should favor the cycloaddition from one face of the molecule to provide cycloadducts with a high degree of diastereofacial control. In the extension of this reaction to glycals, furanoid glycals were initially used as substrates. The study was undertaken with furanoid glycals for two main reasons: 1) furanoid glycals were expected to show a greater reactivity than pyranoid glycals and 2) with furanoid glycals, the C-3 substituent should have a strong directing effect.

Synthesis of Methyl 2-Amino-2-Deoxy-Glycofuranosides

The furanoid glycals (**1-3**) (Table I) used in the present study were efficiently prepared via reductive elimination of 2,3-isopropylidene glycosyl chlorides (*21-22*) or via dehydration of a lactol. These glycals underwent [4+2] cycloaddition reactions with dibenzyl azodicarboxylate under irradiation at 350 nm in cyclohexane. In each case, a single cycloadduct (**4-6**) was isolated in which the newly formed carbon-nitrogen bond at C-2 was shown to be trans to the C-3 substituent. The adducts were subsequently converted, with inversion of configuration at C-1, to the corresponding methyl glycosides (**7-9**) by treatment with methanol and p-toluenesulfonic acid. The use of dibenzyl azodicarboxylate (DBAD) for the cycloaddition reaction allows the transformation of the hydrazides to the corresponding amines (**10-12**) in a single operation under hydrogenolysis conditions (Ra-Ni, MeOH, 40 PSI H_2). The stereochemistry at C-1 and C-2 of the methyl 2-amino-2-deoxy-furanosides was secured by comparison with known compounds (*23-25*).

Synthesis of Methyl 2-Amino-2-Deoxy-Glycopyranosides

In light of the efficiency of the method in the preparation of methyl 2-amino-2-deoxy-furanosides, it was normal to adapt the method for the preparation of 2-amino pyranosides, since amino sugars usually exist as the pyranoside form in natural products. In some cases, 2-amino furanosides have been converted to 2-amino pyranosides under acidic conditions (*26*). For example, the glucofuranoside **15** prepared from glucal **13** was treated with acidic methanol to give, after deprotection and derivation, the tetraacetyl methyl 2-amino-2-deoxy-ß-L-glucopyranoside (**16**) (Figure 2). However, the most attractive and straightforward route for the synthesis of 2-amino-2-deoxy-pyranosides was to perform the cycloaddition reaction on pyranoid glycals.

For this study, the commercially available triacetyl D-glucal (**17**) (Table II) was the first substrate subjected to the cycloaddition conditions. Unfortunately, under the conditions used for furanoid glycals, the cycloaddition did not take place. However, when the reaction was carried out at 90°C a 3/2 diastereomeric mixture of adducts **27** was obtained in low yield. Since dihydropyran was found to be a good substrate in the cycloaddition reaction under irradiation, this suggested that the low reactivity of triacetyl D-glucal was due to the acetoxy groups on the ring.

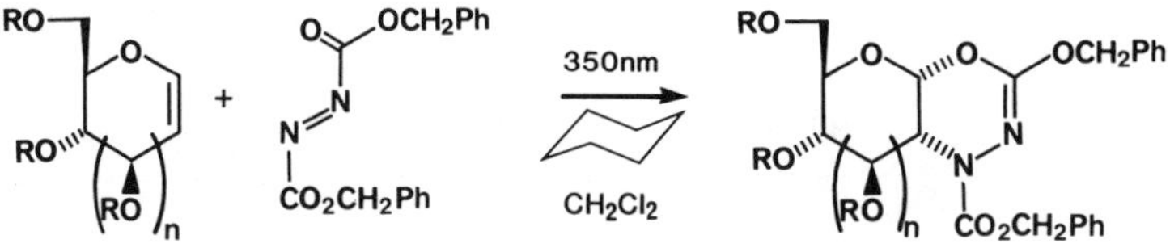

n = 0,1

Figure 1. The (4+2) cycloaddition reaction of dibenzyl azodicarboxylate with glycals.

TABLE I: SYNTHESIS OF METHYL 2-AMINO-2-DEOXY GLYCOFURANOSIDES

GLYCAL	[4+2] ADDUCT	GLYCOSIDE	FREE AMINE
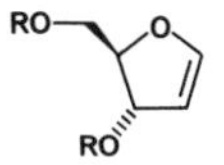 1	4 70%	7 88%	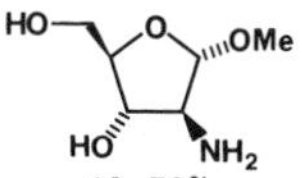 10 73%
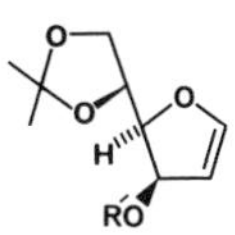 2	5 73%	8 86%	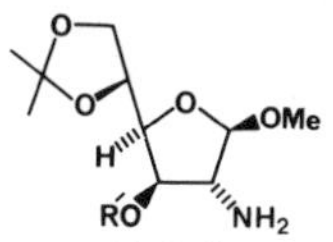 11 96%
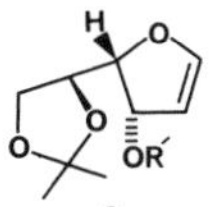 3	6 80%	9 80%	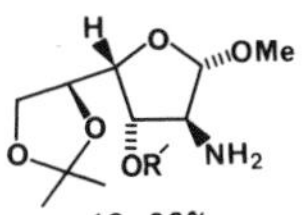 12 98%

R=t-BuPh$_2$Si–
R′=t-BuMe$_2$Si–

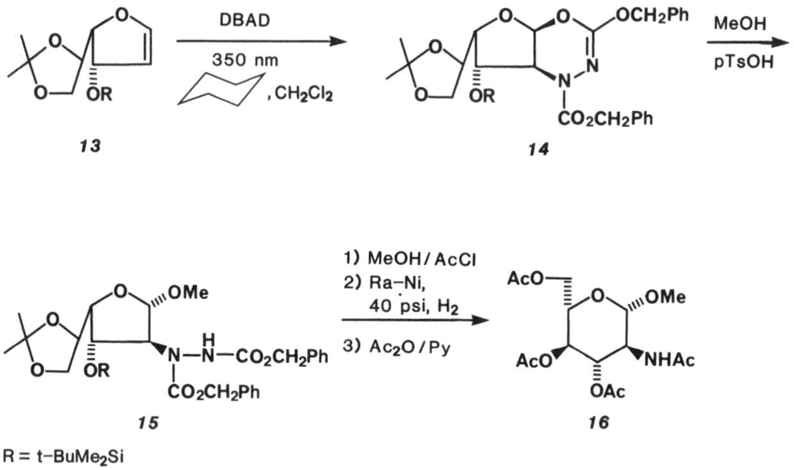
DBAD
350 nm
, CH2Cl2
OR
OCH2Ph
CO2CH2Ph
MeOH
pTsOH
13
14
OMe
N—N—CO2CH2Ph
CO2CH2Ph
1) MeOH / AcCl
2) Ra-Ni,
40 psi, H2
3) Ac2O / Py
AcO
NHAc
OAc
15
16
R = t-BuMe2Si

Figure 2. Synthesis of tetraacetyl methyl 2-amino-2-deoxy-ß-L-glucopyranoside.

TABLE II : SYNTHESIS OF METHYL 2-AMINO-2-DEOXY-GLYCOPYRANOSIDES

SUBSTRATE	ADDUCT	GLYCOSIDE	AMINE	
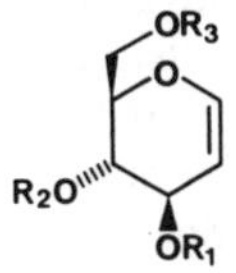			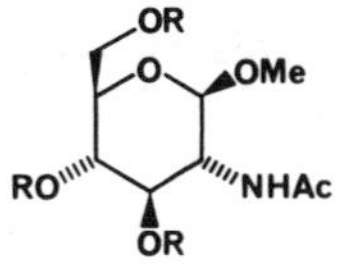	
17 R_1, R_2, R_3=Ac	***27*** 20% (3/2)	—	***38***→***41*** 72%	R = t-BuMe$_2$Si
18 R_1, R_2, R_3=t-BuMe$_2$Si	***28*** 70%	***38*** 88%		
19 R_1, R_2=Ac R_3= t-BuMe$_2$Si	***29*** 0%	—		
20 R_1 R_2=Me R_3=Ac	***30*** 60% (4/1)	—	→ ***42***	R = Ac
21 	***31*** 70%	***39*** 80%	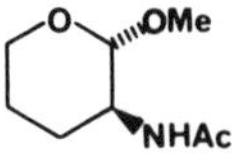 ***39***→***43*** 75%	
22 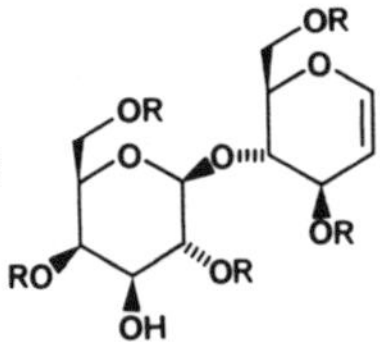 R = t-BuMe$_2$Si	***32*** 94%	—	—	
23 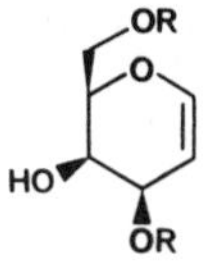 R = t-BuMe$_2$Si	***33*** 86%	—	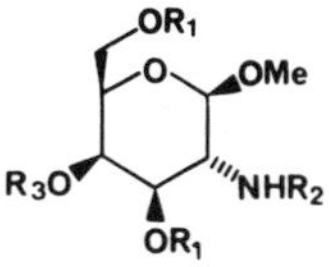 ***33***→***44*** 60%	R_1= t-BuMe$_2$Si R_2, R_3 =H
			→ ***45***	R_1, R_2, R_3 =Ac
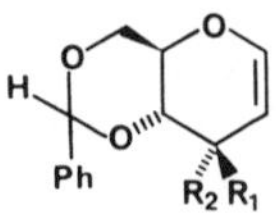				
24 R_1= t-BuMe$_2$SiO R_2=H	***34*** 70% (15/1)	***40*** 82%	***40***→***42***	
25 R_1=H R_2= t-BuMe$_2$SiO	***35*** 70%	—	—	
26 R_1,R_2=H	***36*** 70% 4 TOP	—	—	
	37 1 BOTTOM	—	—	

When the acetyl groups of triacetyl D-glucal were replaced with *tert*-butyldimethylsilyl groups (**18**), the reaction took place efficiently to provide the dihydrooxadiazine **28** in good yield. Interestingly in this case, the diene adds exclusively from the α face of the molecule, anti to the C-3 substituent, even though the glycal would be expected to exist in a half chair conformation in which the C-3 silyloxy group would be in a pseudo equatorial orientation. Ring opening of the adduct **28** with methanol was followed by hydrogenolysis. ^1H NMR analysis of the methyl N-acetyl trisilyl D-glucopyranoside **41** suggested that no trans diaxial couplings were present. This result implies that the molecule exists in a $^1C_4$ or in a boat conformation. This unusual conformation for a glucose derivative is probably due to the disfavoured gauche relationship between the two silyloxy groups at C-3 and C-4 in the $^4C_1$ conformation. For the same reason, the glucal **18** probably exists in a half boat conformation in which the axial C-3 substituent would impede the cycloaddition from occurring from the ß face of the molecule. Similarly, with pentasilyl lactal (**22**) as substrate *(27)*, the reaction took place only from the α face of the molecule, opposite to the C-3 substituent. When disilyl D-galactal (**23**) was subjected to the cycloaddition conditions again a single isomer (**33**) was produced. The adduct was then converted to methyl tetraacetyl 2-amino-2-deoxy-ß-D-galactopyranoside (**45**) using the previously described sequence.

In these previous examples of pyranoid glycals, the orientation of the C-3 substituent influences the reaction in the same sense as in the case of furanoid glycals. However, the orientation at C-3 was not completely fixed due to the conformational flexibility of pyranoid and furanoid glycals. In order to gain a better understanding of the directing effect of the C-3 substituent, the reaction was performed with more rigid glycals. The 4-6-benzylidene glycals **24-26** were prepared *(28-30)* and submitted to the cycloaddition conditions. As anticipated in the case of allal (**25**), the silyloxy at C-3 hinders the α face of the molecule and thus the cycloaddition took place exclusively from the ß face of the molecule. With the 4,6-benzylidene-D-glucal **24**, the selectivity was reversed as a 15/1 diastereomeric mixture was produced, the major isomer corresponding to an addition from the α face of the molecule, again anti to the C-3 substituent. With the deoxy case (**26**), little selectivity was expected due both to the remoteness of the chiral centers from the addition point and the general planarity of the molecule. To our surprise, a 4/1 mixture of isomers was produced, the major one being the result of an addition from the ß face of the molecule. The formation of an axial carbon-nitrogen bond at C-2 giving rise to a quasi-chair conformation in the transition state of the reaction might explain this selectivity. In order to confirm that this result was not a consequence of a thermodynamic equilibration, the two isomers were separately submitted to the reaction conditions. Neither the glycal nor the corresponding isomer were produced.

Synthesis of Complex 2-Amino Saccharides

It had been demonstrated hitherto that the cycloaddition reaction is equally efficient with five and six membered glycals and that the C-3 substituent has in both cases a strong directing effect on the cycloaddition. This selectivity encountered in the introduction of an amino function at C-2 of a carbohydrate combined with the incorporation of a latent nucleofugal group at C-1, both via the cycloaddition, made the method very attractive for the preparation of complex 2-amino saccharides. As described above, the dihydrooxadiazines obtained from furanoid and pyranoid glycals can be efficiently opened by methanol under acidic conditions to provide the methyl 2-amino-2-deoxy-glycosides with complete inversion at C-1. By analogy, the adducts would be expected to undergo ring opening with more elaborate alcohols to give an entry into the synthesis of complex 2-amino saccharides. In the previous examples, methanol was the solvent of the reaction. Thus it was not obvious that the ring opening would occur using a stoichiometric amount of a primary or a secondary alcohol in an alternative solvent. The first attempt was aimed at the synthesis of the C-1 C-6 disaccharide **47** (Figure 3). This was easily accomplished by treating the adduct **28**, prepared from trisilyl D-glucal, with 1.2 equivalent of diacetonide D-galactose (**46**) in CH_2Cl_2 at 25°C in the presence of p-toluenesulfonic acid. The disaccharide **47** was isolated in 95% yield as a single isomer, and the free amine **48** was generated under the same conditions used for the simple cases described above.

Following this promising result, we focused our efforts on a more demanding case, the coupling of the adduct **28** with the free C-3 hydroxy group of the glucose derivative **49**. Unfortunately, the use of p-toluenesulfonic acid as the catalyst yielded a complex mixture of products. This is probably due to the poor reactivity of the adduct towards a sterically hindered alcohol, resulting in the degradation of both starting materials and coupled product under the acidic reaction conditions. In an attempt to overcome these difficulties, $BF_3{\bullet}Et_2O$ was used as an activator of the dihydrooxyadiazine **28**. To our surprise, the 1,6-anhydro sugar **50** was produced rather than the expected intermolecular coupling product. The structure of the 1,6-anhydro sugar was confirmed by comparison of the physical constants and ^1H NMR data of the triacetyl derivative **51** with those reported (*31*). In order to make the C-6 oxygen less nucleophilic, the triacetyl D-glucal adduct **27** was prepared from **28** and subjected to the glycosidation conditions. With $BF_3{\bullet}Et_2O$ as the catalyst, the C-1 C-3 disaccharide **52** was produced in good yield without any formation of 1,6-anhydro sugar. Encouraged by this result, we decided to apply the method to the synthesis of a C-1 C-1 2-amino disaccharide, since some antibiotics such as tunicamycin contain this linkage. Our method had actually been used in the synthesis of protected subunits of tunicamycines (*32*), but the direct formation of the ß,α-trehalose bond involving the dihydrooxadiazine as glycosylating agent had never been reported. At this stage, several attempts were made to open a dibenzyl adduct with a lactol but without any success. As

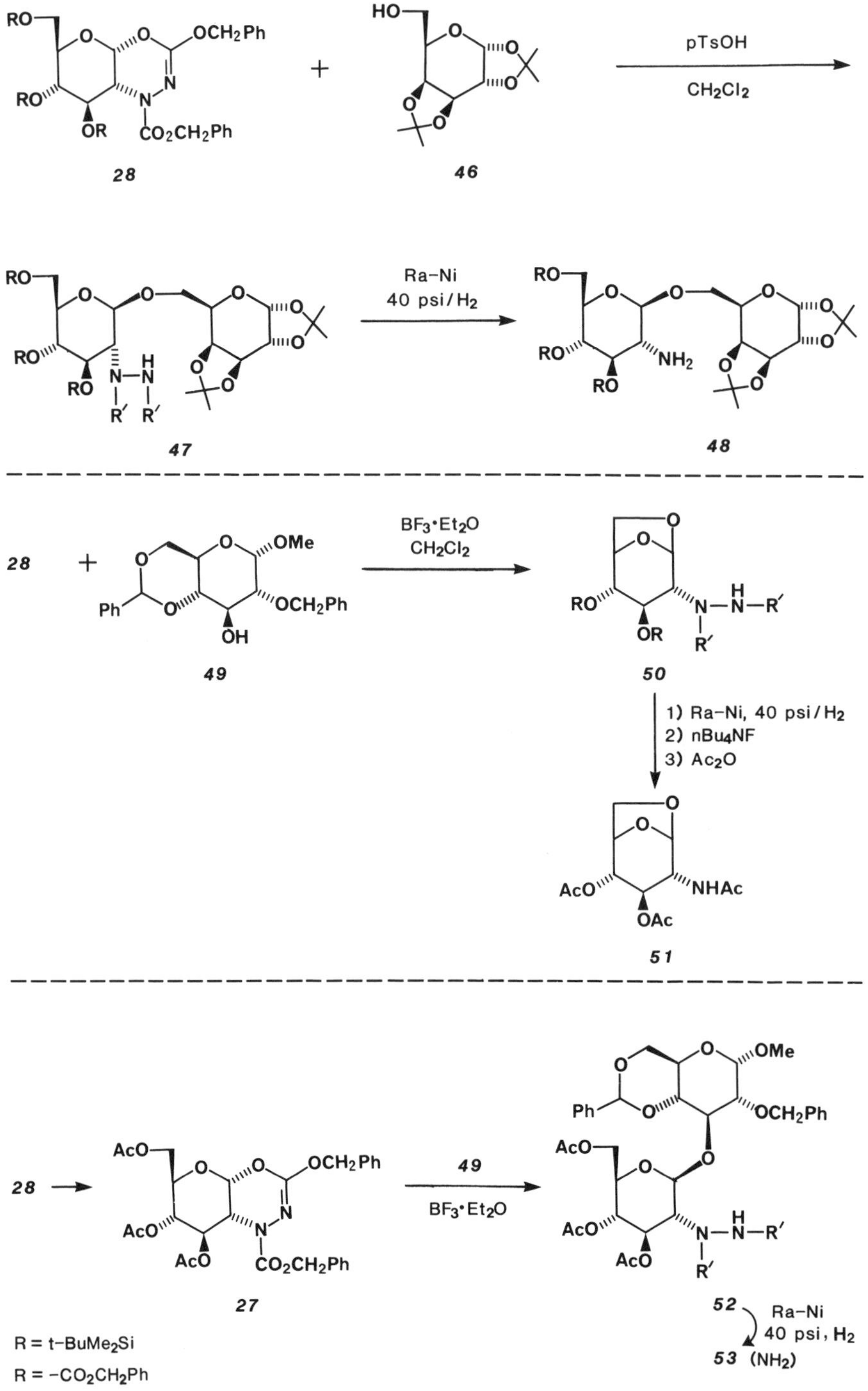

RO OCH2Ph N N CO2CH2Ph RO OR 28 + HO 46 pTsOH CH_2Cl_2

47 Ra–Ni 40 psi/H_2 48 NH2

28 + 49 OMe OCH2Ph Ph OH $BF_3 \cdot Et_2O$ CH_2Cl_2 50

1) Ra–Ni, 40 psi/H_2
2) nBu_4NF
3) Ac_2O

51 AcO NHAc OAc

28 → 27 AcO OCH2Ph CO2CH2Ph 49 $BF_3 \cdot Et_2O$ 52 Ra–Ni 40 psi, H_2 53 (NH_2)

R = t-BuMe2Si
R = $-CO_2CH_2Ph$

Figure 3. Synthesis of complex 2-amino saccharides.

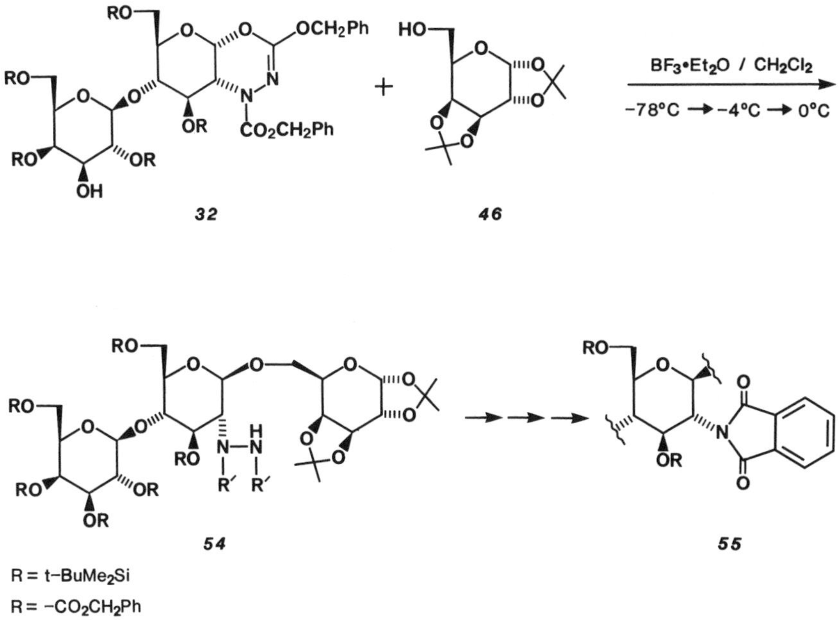

R = t-BuMe$_2$Si
R = -CO$_2$CH$_2$Ph

Figure 4. Preparation of a 2-amino trisaccharide.

we will see below, this problem has been partially solved by changing the ester groups on the dihydrooxadiazines.

The method can also be applied to the synthesis of 2-aminotrisaccharides. For example when the D-lactal adduct **32** was treated with diacetonide galactose (**46**) in CH_2Cl_2 at high concentration to avoid the formation of the 1,6-anhydro sugar, the trisaccharide **54** was obtained in 82% yield (Figure 4). After deprotection and derivatization, the trisaccharide **55** was obtained which is an analogue of blood group determinant previously prepared (*33*).

Having shown the usefulness of the method for the synthesis of oligosaccharides, we then turned our efforts to the opening of the adducts with other types of nucleophiles. For example, the dihydrooxadiazine **28** was subjected to ring opening by dihydrocholesterol to provide, within few minutes, the saccharide **58** in 96% yield, with complete inversion at C-1 (Figure 5). We also looked at the possibility of introducing a labile protecting group at C-1, which could be eventually removed under mild conditions later in a synthetic scheme. In this regard, the adduct **28** was opened with acetic acid to introduce an acetoxy group at C-1 (**56**) which is sufficiently stable to not be altered under the hydrogenolysis conditions. The same adduct could also be opened with hydride using $NaBH_3CN$ and ZnI_2 to afford the 1,5-anhydro sugar **60**. Repeating the experiment this time with $NaBD_3CN$, the deuterium was introduced in the ß-orientation for at least 90% of the product, based on ^1H NMR analysis, demonstrating the S_N2 character of the reaction.

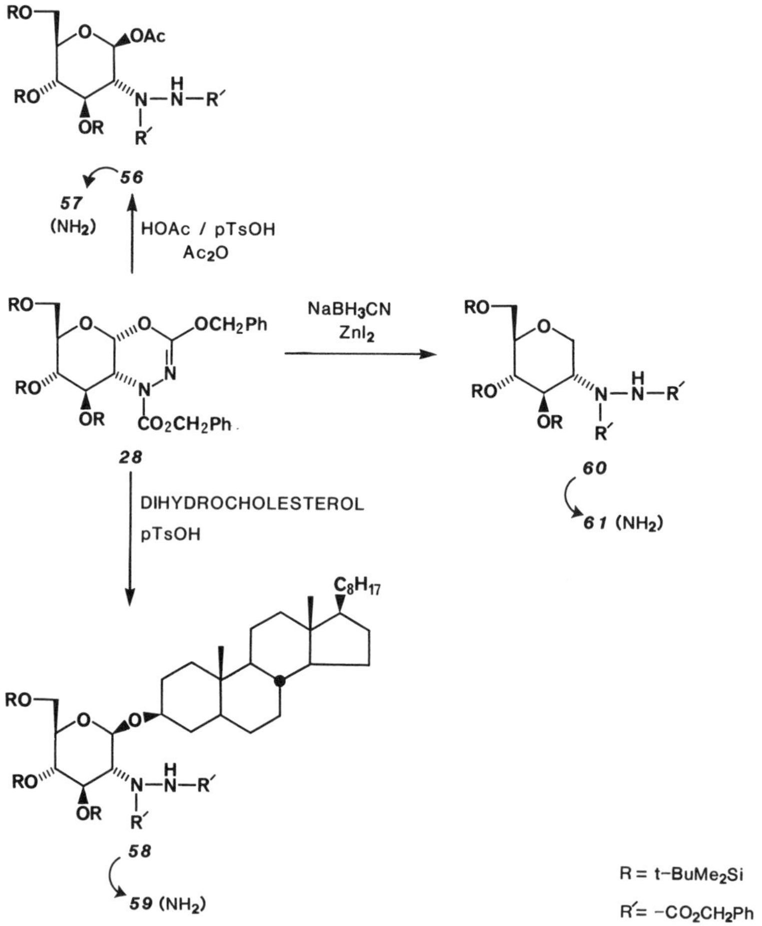

Figure 5. Opening of adduct **28** with various nucleophiles.

Inhibitory Effect of the Acyloxy Group

After the scope and limitations of our method had been well established, we were still intrigued by the inhibitory effect of acetoxy units on the cycloaddition reaction, as observed with triacetyl D-glucal. The acetoxy groups at C-3 and C-6 were suspected of inhibiting the reaction and therefore the two glucal **19** and **20** (Table II) were prepared and subjected to the cycloaddition conditions. With glucal **19** having an acetoxy group at C-3 and a silyloxy group at C-6, no cycloadduct was produced after several days of irradiation in the presence of dibenzyl azodicarboxylate. However, with a methyl ether at C-3 and an acetoxy at C-6, the reaction took place smoothly to produce a 4/1 diastereomeric mixture of the adduct **30**. From these findings, it was clear that an electron withdrawing group at C-3 decreases the reactivity of vinyl ethers towards the azo compound. This inhibitory effect of an electron withdrawing group is probably the result of a lowering in the HOMO energy of the vinyl ethers by the allylic acyloxy group, affecting the overlap with the LUMO of the azo compound. In contrast, the lowering of the LUMO energy of the azo compound by the introduction of electron withdrawing groups should also increase the rate of the cycloaddition. This concept lead us to improve on the cycloaddition reaction by modifying the azo compound.

In the process of selecting a new azo compound to perform the cycloaddition reaction, we considered that the removal of the hydrazide esters should be accomplished under reductive conditions to allow simultaneous cleavage of the nitrogen-nitrogen bond and thus avoid multiple steps in the preparation of the free amine. It seemed that, bis(2,2,2-trichloroethyl) azodicarboxylate (BTCEAD) fulfilled these criteria since it contains electron withdrawing groups which could be removed under reductive conditions. This reagent was initially prepared as described in the literature *(34)* and placed under the cycloaddition conditions with glycals (Table III). It was noticed that this azo compound adds to glycals at higher rates than DBAD *(35)*. Even with triacetyl D-glucal as substrate (**17**), the cycloaddition reaction took place under irradiation conditions, albeit at a low rate, to provide a 2/3 diastereomeric mixture of adduct **66**. Triacetyl D-galactal **62**, which also bears an electron withdrawing group at C-3, underwent cycloaddition with BTCEAD to provide a single cycloadduct (**68**). The usefulness of this reagent is also apparent with glycals having no electron withdrawing group attached at C-3, since the reaction time and the number of equivalents of the azo compound have been considerably reduced. As observed in the benzyl series, the trichloroethyl adducts (**64-69**) can be opened by simple alcohols such as methanol to yield the methyl glycosides (**70-74**).

In addition to being more reactive than DBAD in the cycloaddition reaction, other advantages were found in the use of BTCEAD. As mentioned earlier, no success was obtained while attempting to prepare a C-1 C-1 2-amino disaccharide by opening a dibenzyl adduct with a lactol. The bis(trichloroethyl) adducts, on the other hand, can easily be converted to these structurally complex C-1 C-1 2-amino disaccharides. For example (Table III), the C-1 C-1 disaccharide **75** was produced by condensation of L-diacetonide mannose on adduct **69** using $BF_3 \cdot Et_2O$ as a catalyst. It is noteworthy that the disaccharide **75** was isolated as a single isomer at both anomeric centers.

The next step was to find conditions for the transformation of the trichloroethyl hydrazides to their corresponding free amines. Under catalytic hydrogenation with Ra-Ni, the amines were isolated only in low yields. Zinc dust in acetic acid was not more successful since the hydrazino sugars were isolated instead of the desired amino sugars. It was then found that addition of

TABLE III : AMINATION OF GLYCALS WITH BIS(2,2,2-TRICHLOROETHYL) AZODICARBOXYLATE

GLYCAL	ADDUCT	GLYCOSIDE	AMINE
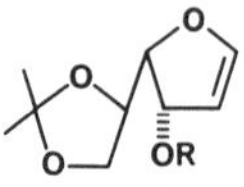 O, O, O, OR *13*	*64* 73% 18 hr	*70* 95%	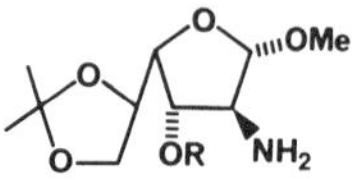 O, O, O, OMe, OR, NH_2 *76* 65%
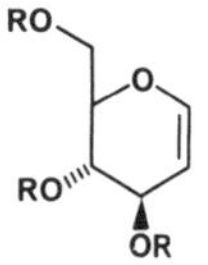 RO, O, RO, OR *18*	*65* 80% 1.5 days	*71* 88%	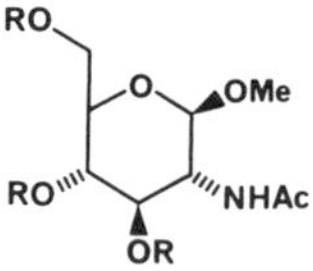 RO, O, OMe, RO, NHAc, OR *41* 80%
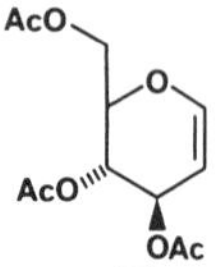 AcO, O, AcO, OAc *17*	*66* 76% 2 weeks 2/3 Top/Bottom	—	—
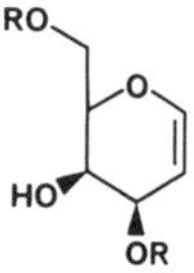 RO, O, HO, OR *23*	*67* 78% 1 day	*72* 75%	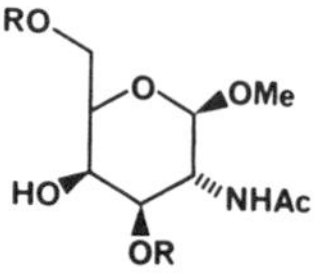 RO, O, OMe, HO, NHAc, OR *44* 70%
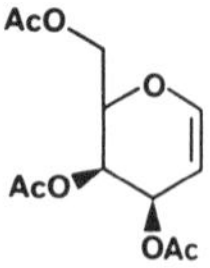 AcO, O, AcO, OAc *62*	*68* 85% 1.5 weeks	*73* 97%	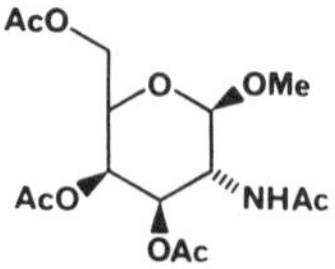 AcO, O, OMe, AcO, NHAc, OAc *45* 75%
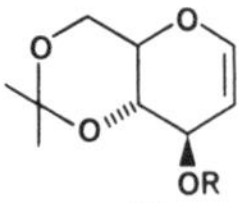 O, O, O, OR *63*	*69* 80% 5 days 15/1 Top/Bottom	*74* 90%	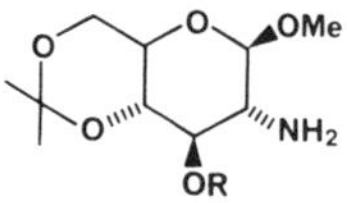 O, O, OMe, O, NH_2, OR *77* 90%
		75 60-68%	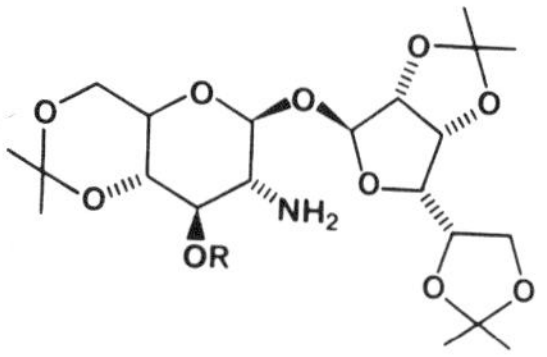 O, O, O, O, NH_2, OR, O, O, O, O, O *78* 90%

R=t-BuMe$_2$Si

SOURCE : ADAPTED FROM REF. 35

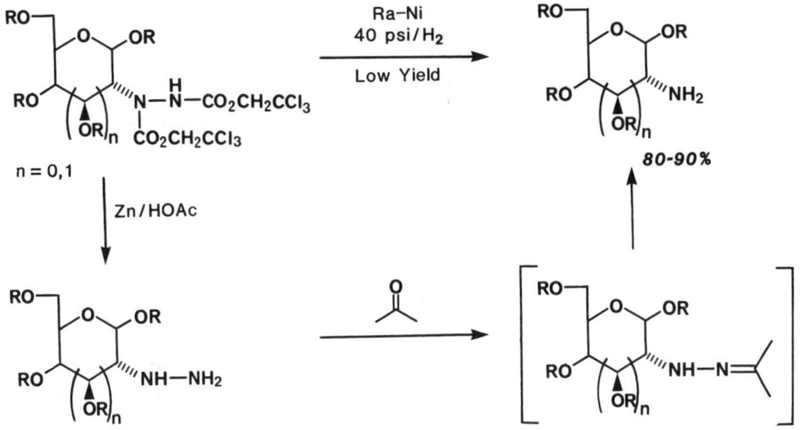

Figure 6. Transformation of trichloroethyl hydrazides to free amines Source: Adapted from Ref. 35.

acetone to this reaction mixture lead to the formation of the amines in good yield. The transformation of the hydrazine to the amine was shown to proceed via the hydrazone (Figure 6). A hydrazone intermediate has actually been isolated and placed under the experimental conditions without acetone to produce the amine, showing that it is an intermediate in the formation of the amine. The zinc/acetic acid/acetone conditions were also applied to the C-1 C-1 disaccharide **75** to yield the free amine **78** in 90%. The stereochemistry at both anomeric centers was proven by ^{1}H NMR analysis (Figure 7).

Based on these results, it is obvious that BTCEAD has a broader spectrum of use than DBAD. In the future, this reagent could be particularly useful for molecules containing functionalities such as olefins or benzyl groups, where hydrogenolysis conditions are proscribed.

In summary, dihydrooxadiazines were efficiently prepared by the hetero [4+2] cycloaddition reaction of azodicarboxylates and glycals. These dihydrooxadiazines can be converted to 2-amino saccharides. This method is presently being extended to the synthesis of 2-amino C-glycosides and results will be reported in a forthcoming paper.

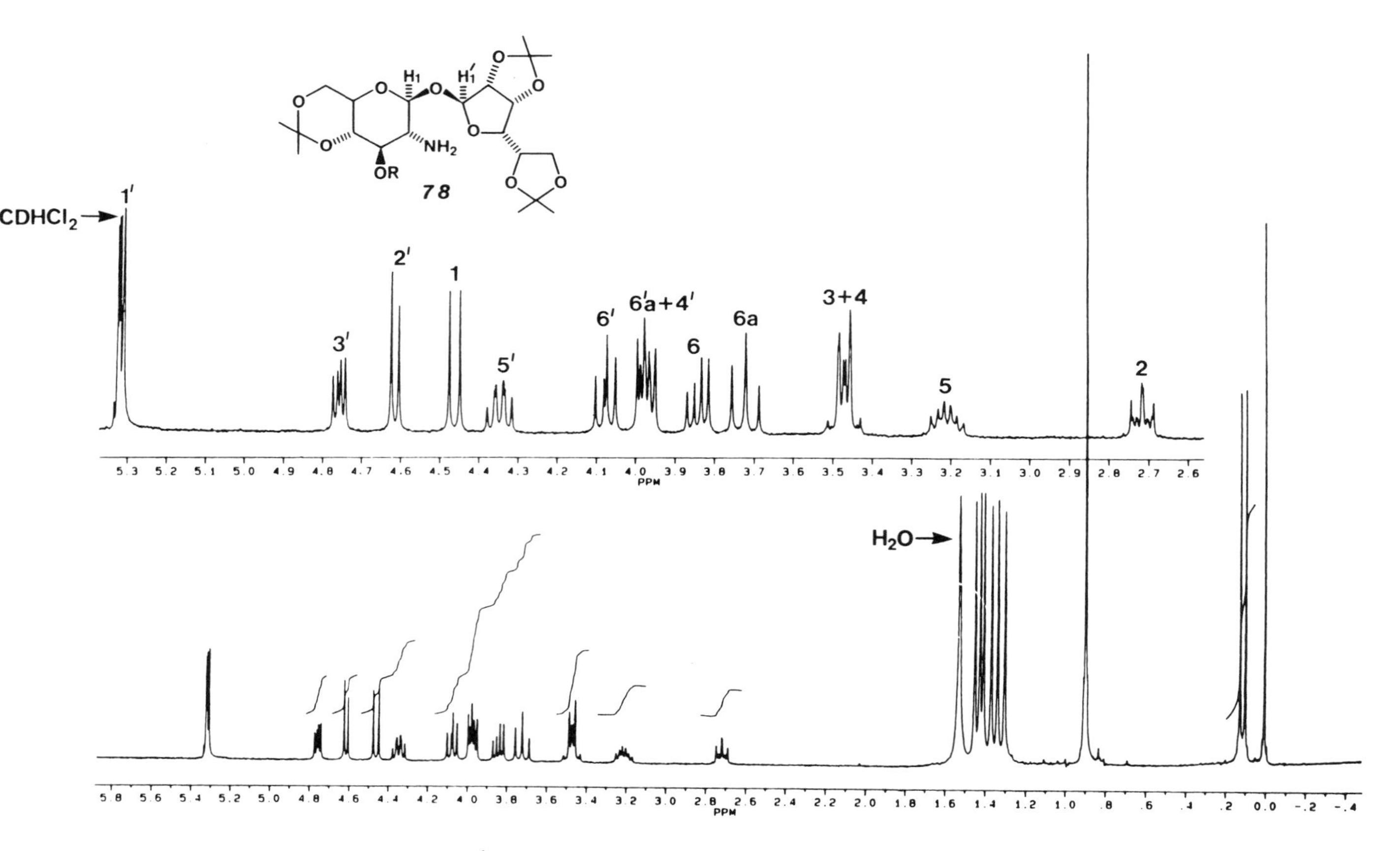

Figure 7. ^{1}H NMR spectrum (300 MHz) of disaccharide **78**.

Acknowledgments

The authors wish to thank M. Bernstein for performing elevated-temperature NMR spectra of the 2-hydrazido sugars, M. Wacasey for information support, Marcy Mintz for proof reading and T. Grassetti for the typing of this chapter.

Literature Cited

1. Watkins, W.M.; Greenwell, P.; Yates, A.D.; Johnson, *P.H. Biochimie*, **1988**, *70*, 1597-1611.
2. Pan, Y.T.; Elbein, A.D. *Prog. Drug Res.*, **1990**, *34*, 163-207.
3. Tkacz, J.S. *Antibiotics*, **1983**, *6*, 255-278.
4. Edson, R.S.; Terrell, C.L. *Mayo Clin. Proc.*, **1987**, *62*, 916-920.
5. Hajivarnava, G.S.; Overend, W.G.; Williams, N.R. *J. Chem. Soc., Perkin Trans 1*, **1982**, 205-214.
6. Dyong, I.; Schulte, G. *Tetrahedron Lett.*, **1980**, *21*, 603-606.
7. Lemieux, R.U., Nagabhushan, T.L. *Can. J. Chem.* **1968**, *46*, 401-403.
8. Lemieux, R.U., Ratcliffe, R.M. *Can. J. Chem.* **1979**, *57*, 1244-1251.
9. Griffith, D.A.; Danishefsky, S. *J. Am. Chem. Soc.*, **1990**, *112*, 5811-5819.
10. Leblanc, Y.; Fitzsimmons, B.J.; Springer, J.P.; Rokach, J. *J. Am. Chem. Soc.*, **1989**, *111*, 2995-3000.
11. Fitzsimmons, B.J.; Leblanc, Y.; Chan, N.; Rokach, J. *J. Am. Chem. Soc.* **1988**, *110*, 5229-5231.
12. Fitzsimmons, B.J.; Leblanc, Y.; Rokach, J. *J. Am. Chem. Soc.*, **1987**, *109*, 285-286.
13. Firl, J.; Sommer, S. *Tetrahedron Lett.* **1972**, 4713-4716.
14. Firl, J.; Sommer, S. *Tetrahedron Lett.* **1971**, 4193-4196.
15. Firl, J.; Sommer, S. *Tetrahedron Lett.* **1970**, 1929-1932.
16. Koerner Von Gustorf, E.; White, D.V.; Kim, B.; Hess, D.; Leitich, J. *J. Org. Chem.* **1970**, *35*, 1155-1165.
17. Firl, J.; Sommer, S. *Tetrahedron Lett.* **1970**, 1925-1928.
18. Koerner Von Gustorf, E.; White, D.V.; Leitich, J.; Henneberg, D. *Tetrahedron Lett.*, **1969**, 3113-3116.
19. Firl, J.; Sommer, S. *Tetrahedron Lett.* **1969**, 1137-1140.
20. Firl, J.; Sommer, S. *Tetrahedron Lett.* **1969**, 1133-1136.
21. Ireland, R.E.; Thaisrivongs, S.; Vanier, N.; Wilcox, C.S. *J. Org. Chem.*, **1980**, *45*, 48-61.
22. Ireland, R.E.; Wilcox, C.S.; Thaisrivongs, S. *J. Org. Chem.* **1978**, *43*, 786-787.
23. Jacquinet, J.C.; Sinay, P. *Carbohydr. Res.* **1974**, *32*, 101-114.
24. Montgomery, J.A.; Thorpe, M.C.; Clayton, S.D.; Thomas, H.J. *Carbohydr. Res.* **1974**, *32*, 404-407.

25. Buchanan, J.G.; Clark, D.R. *Carbohydr. Res.* **1977**, *57*, 85-96.
26. Konstas, S.; Photaki, I.; Zervas, L. *Chem. Ber.* **1959**, *92*, 1288-1293.
27. Haworth, W.H.; Hirst, E.L.; Plant, M.M.T.; Reynolds, R.J.W. *J. Chem. Soc.* **1930**, 2644-2653.
28. Sharma, M.; Brown, R.K. *Can. J. Chem.* **1966**, *44*, 2825-2835.
29. Lemieux, R.V.; Fraga, E.; Watanabe, K.A. *Can. J. Chem.* **1968**, *46*, 61-69.
30. Fraser-Reid, B.; Radatus, B. *J. Am. Chem. Soc.* **1970**, *92*, 6661-6663.
31. Schmitt, F.; Sinay, P. *Carbohydr. Res.* **1973**, *29*, 99-111.
32. Danishefsky, S.; DeNinno, S.L.; Chen, S.H.; Boisvert, L.; Barba-Chyn, M. *J. Am. Chem. Soc.* **1989**, *111*, 5810-5818.
33. Lemieux, R.U.; Abbas, S.Z.; Chung, B.Y. *Can. J. Chem.* **1982**, *60*, 68-75.
34. Mackay, D.; Pilger, C.W.; Wong, L.L. *J. Org. Chem.* **1973**, *38*, 2043-2049.
35. Leblanc, Y.; Fitzsimmons, B.J. *Tetrahedron Lett.* **1989**, *30*, 2889-2892.

RECEIVED December 2, 1991

Chapter 7

Glycosylmanganese Pentacarbonyl Complexes

An Organomanganese-Based Approach to the Synthesis of C-Glycosyl Derivatives

Philip DeShong, Greg A. Slough, D. Richard Sidler, Varadaraj Elango, Philip J. Rybczynski, Laura J. Smith, Thomas A. Lessen, Thuy X. Le, and Gary B. Anderson

Department of Chemistry and Biochemistry, University of Maryland, College Park, MD 20742

Preparation and reactions of glycosylmanganese pentacarbonyl complexes are discussed. Anomerically pure complexes of pyranosyl- and furanosyl-complexes have been prepared in excellent yield and shown to undergo various migratory insertion reactions resulting in formation of C-glycosyl derivatives. Applications of this technology to the synthesis of C-glycosyl and C-aryl glycosidic systems is discussed.

C-Glycosyl compounds are an interesting class of compounds due to their ability to function as nucleoside surrogates. As such, both naturally occurring and synthetic C-glycosyl derivatives have served as biochemical probes*(1)* and have been shown to function as antibiotics*(2)*, antitumor and antiviral agents*(3)*. For example, the naturally occurring C-glycosyls vineomycin (**1**)*(4)*, gilvocarcin (**2**) *(5)*, and pyrazofurin (**3**)*(6)* possess potent biological activity. Similarly, tiazofurin (**4**) and its synthetic analogue selenazofurin (**5**) are in clinical trials as antitumor agents *(7)*.

The goal of this project has been to develop a general approach to the synthesis of C-glycosyl derivatives based upon an extension of our previous studies into the chemistry of alkylmanganese pentacarbonyl complexes *(8-16)*. As summarized in Scheme 1, it has been demonstrated that simple alkyl complexes such as methylmanganese pentacarbonyl (**6**) can be transformed into a variety of carbonyl derivatives with formation of either one, two, or three carbon-carbon bonds *(9, 10, 12)*.

In the most common reaction of alkylmanganese complexes, ester, amide, and thioester derivatives (**8**) are prepared in excellent overall yield by the Reppe reaction. Migratory insertion of carbon monoxide affords acyl complex **7** *with formation of a new carbon-carbon bond.* Cleavage of the acyl-metal bond by alcohols, amines, or thiols in the presence of Na_2CO_3 furnishes the respective carboxylic acid derivatives. This transformation, although well studied in the organometallic literature, will furnish new insights into the migratory insertion process when applied to complex glycosyl systems (*vide infra*).

0097–6156/92/0494–0097$06.00/0

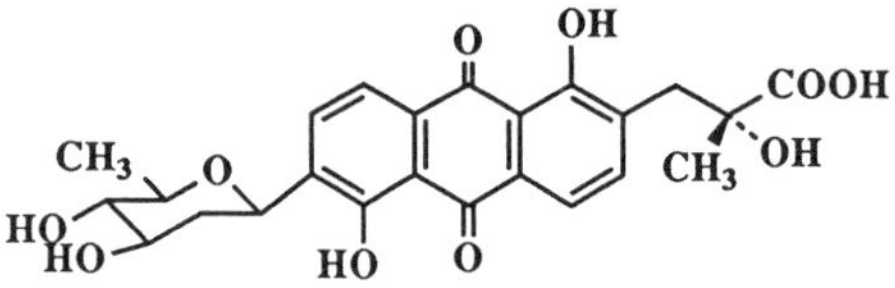

Vineomycinone B_2 (**1**)

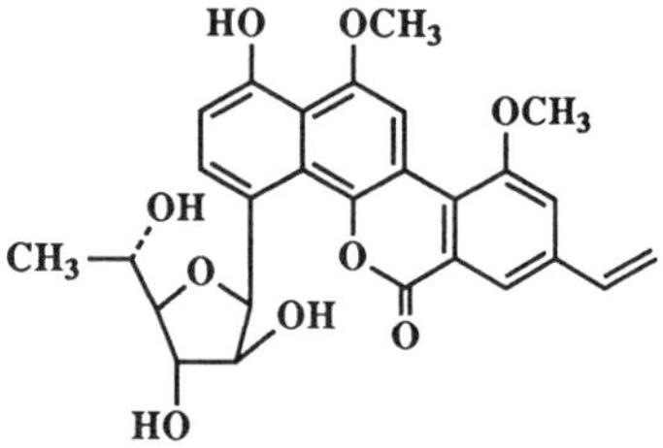

Gilvocarcin V (**2**)

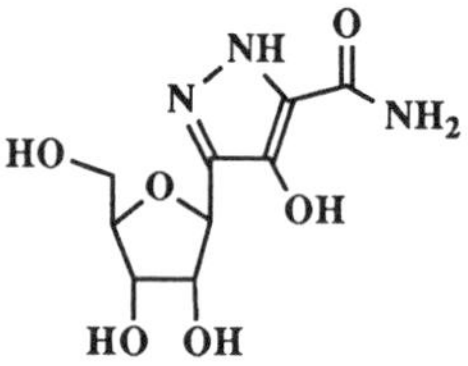

Pyrazofurin (**3**)

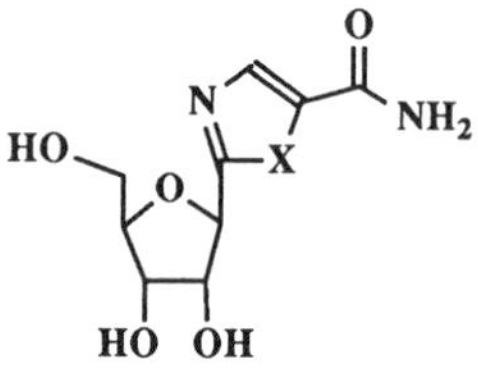

Tiazofurin (**4**) : X = S
Selenazofurin (**5**) : X = Se

Scheme 1

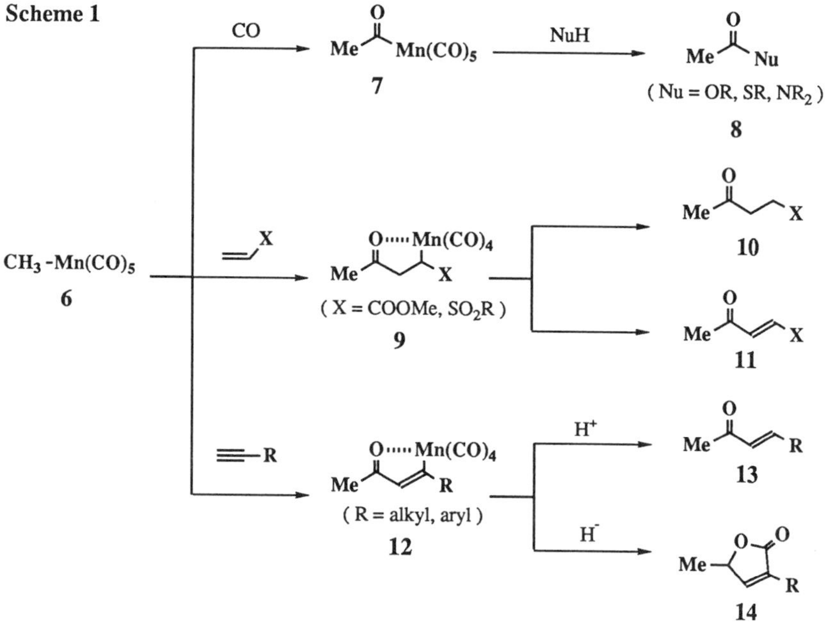

CH_3-$Mn(CO)_5$ **6**

CO → Me(C=O)$Mn(CO)_5$ **7** —NuH→ Me(C=O)Nu (Nu = OR, SR, NR_2) **8**

X-substituted alkene → O·····$Mn(CO)_4$ complex **9** (X = COOMe, SO_2R) → **10**, **11**

alkyne ≡—R → O·····$Mn(CO)_4$ complex **12** (R = alkyl, aryl) —H^+→ **13**; —H^-→ **14**

Sequential insertion of carbon monoxide and an electron-deficient alkene provides manganacycle **9**.Demetallation of complex **9** can be achieved either in a "reductive" manner to furnish ketone **10** or to yield enone **11** with regeneration of the alkene moiety. The overall transformation occurs with formation of two carbon-carbon bonds, incorporation of one molecule of CO, and regioselective functionalization of the alkene substrate.

Finally, sequential insertion of an alkyne with methylmanganese pentacarbonyl gives the unsaturated manganacycle **12**. Two modes of demetalation of **12** have also been developed for this complex. In the first, protonation of the manganacycle provides enone **13**. This reaction has been studied mechanistically in our laboratory *(17)*.

Alternatively, hydride reduction of manganacycle **12** results in the incorporation of a second molecule of carbon monoxide and formation of butenolide **14**. This unique transformation involves formation of three carbon-carbon bonds, two molecules of CO being incorporated, and difunctionalization of an alkyne in a completely regioselective manner.

Having established the versatility of alkylmanganese pentacarbonyls for the preparation of carbonyl derivatives, it was our intention to extend these transformations to glycosylmanganese complexes as a means for preparing C-glycosyl derivatives (See Scheme 2). Successful extension of the manganese methodology would require the investigation of several new aspects of this chemistry. First, it had to be demonstrated that glycosyl derivatives (**16**) could be prepared. With the exception of Trainor's glucopyranosyl Fp-derivative *(18)*, stable transition metal derivatives of carbohydrates had not been reported prior to our studies.

There were stereochemical features to be considered also. Glycosyl complex **16** could exist as either the α- or β-anomer, and it was our intention to synthesize **16** from the protected sugar derivative **15** in a highly stereoselective manner. This is a particularly important point because migratory and sequential insertion processes to give **17-19**, respectively, occur *with retention of configuration (19)*. Accordingly, the configuration of glycosyl complex **16** will be transferred to the C-glycosyl derivatives which result.

Preparation of Pyranosyl- and Furanosylmanganese Pentacarbonyl Complexes

Pyranosyl and furanosyl bromides (and chlorides to a lesser extent) react with alkali metal salts of manganate anion to yield the corresponding glycosylmanganese pentacarbonyl complexes. In many instances, the condensation proceeds in excellent yield and is highly stereoselective. For example, α-glucopyranosyl bromide **20** and potassium manganate (**21**) react at -78 ^{o}C to give *exclusively* β-complex **22** in virtually quantitative yield (Scheme 3).[20] On the other hand, a mixture of α-complex **24** and β-complex **22** results from reaction of bromide **20** and sodium manganate (**23**) in the presence of tetrabutylammonium bromide at -78 ^{o}C *(11, 20)*.

Condensation of sodium manganate **23** with bromide **20** is extremely sensitive to the purity of reagents and the ratios of anomeric products obtained (**22** vs **24**) vary accordingly. Generally, α-anomer **24** predominates *(11)*. We attribute the change in selectivity of the condensation to *in situ* anomerization of the bromide. As the condensation proceeds, bromide ion is liberated and reacts with **20** which results in formation of β-bromide. When the rate of condensation is rapid, as with potassium salt **21**, condensation proceeds more rapidly than anomerization and only β-complex is obtained. However, when the condensation rate is slower, as with **23**, anomerization occurs at a competitive rate to condensation, and mixtures of anomers are obtained. The α-anomer predominates under these conditions presumably

Scheme 2

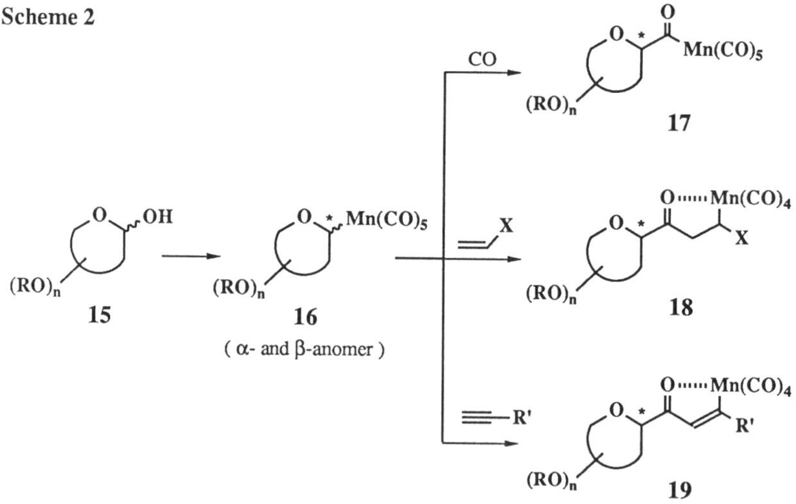

CO

$Mn(CO)_5$

$(RO)_n$

17

O·····$Mn(CO)_4$

X

15

16

(α- and β-anomer)

18

R'

19

Scheme 3

$KMn(CO)_5$

21

THF, -78 °C
>95 %

$Mn(CO)_5$

22

20

$NaMn(CO)_5$

23

n-Bu_4NBr,
THF, -78 °C
75 %

24

+ 22

(24/22 = 3 : 2)

21 or 23 → No Reaction

Me_3Si-$Mn(CO)_5$ → No Reaction

Me_3Si-$Mn(CO)_5$ → No Reaction

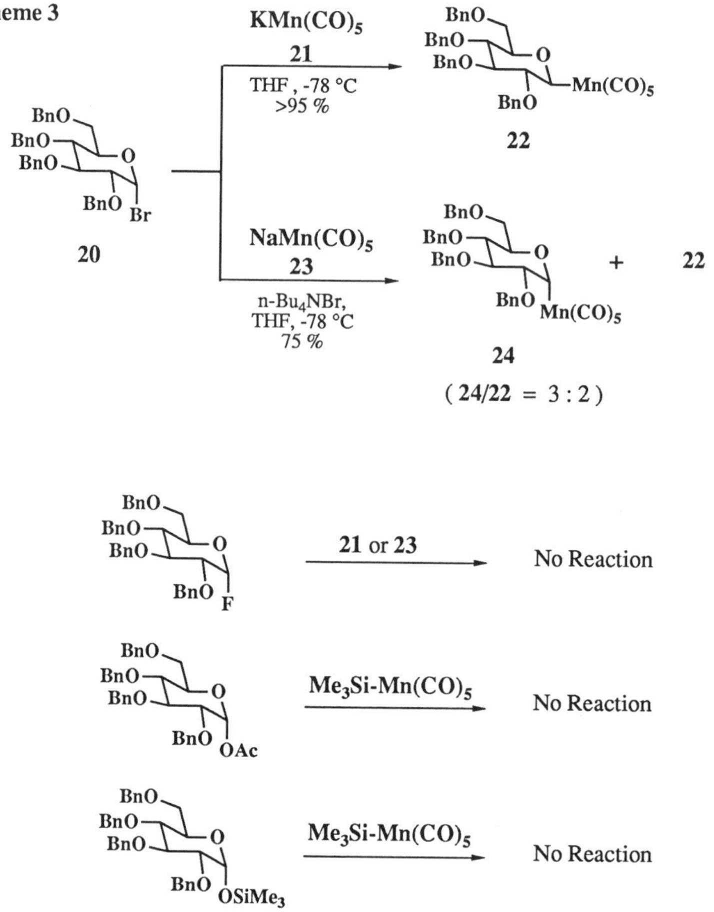

because β-bromide is more reactive than α-bromide **20** *(21)*. It is hoped that optimization studies will provide a protocol such that α-complex **24** can also be prepared from **20** in a highly stereoselective manner.

Attempts at the synthesis of glucopyranosyl complex(es) using the anomeric fluoride, acetate, or silyloxy derivatives, respectively, have not been successful *(20)*.

Chromatographic separation of the α-and β-glucopyranosyl complexes **24** and **22**, respectively, cannot be accomplished. However, a chemical discrimination has been developed. The α-complex **24** undergoes migratory insertion of carbon monoxide approximately seven times faster than β-complex **22** *(22)*. At 40 psi of carbon monoxide, it is possible to convert α-complex **24** to its acyl derivative **25** without inducing formation of the β-acyl complex (Scheme 4). Chromatographic separation of α-acyl **25** and β-glycosyl complex **22** is readily achieved . Gently warming the α-acyl complex induces deinsertion of CO and regenerates α-glucopyranosylmanganese pentacarbonyl (**24**).

Employing condensation conditions similar to those developed for the glucopy-ranosyl system, manganese pentacarbonyl complexes of the α-and β-mannopyranosyl (**26**), β-galactopyranosyl (**27**), α-and β-arabinofuranosyl (**28**), and β-ribofuranosyl (**29**) systems have been synthesized *(8, 11, 20, 23)*.

Anomeric complexes of mannopyranosyl-(**26**) and arabinofuranosylmanganese pentacarbonyl (**28**) are separated in an analogous fashion to the glucopyranosyl analogs. In these instances, *the relative rates of migratory insertion of α/β are* $>10^2$ *and* $>10^3$, respectively *(20)*. This remarkable difference in the rates of migratory insertion in these systems is attributed to the orientation of the carbon-metal bond with regard to the lone pairs on oxygen (an anomeric effect?!) and may provide insight into the factors which control migratory insertion processes. We are currently investigating this phenomenon with other analogs.

Reactions of Glycosylmanganese Pentacarbonyl Complexes

Having prepared glycosylmanganese pentacarbonyl complexes, we turned our attention to demonstrating that these complexes would undergo the processes indicated in Scheme 1. For the most part, the β-glucopyranosyl complex **22** has been employed for these studies. When appropriate, an analogous reaction of the anomeric α-complex **24** has been performed to demonstrate that the migratory insertion reaction occurs with retention of configuration at the anomeric center *(8, 11, 20, 23)*. *For the purpose of this article, only reactions of β-complex **22** are discussed.*

Reppe reaction of complex **22** in the presence of alcohols, amines, or thiols occurs in excellent yield and complete stereoselectivity to afford the corresponding carboxylic acid derivatives (Scheme 5). In the case of β-ribofuranosyl complex **29**, formation of ester **35** constitutes a formal total synthesis of tiazofurin (**4**) and selenazofurin (**5**) *(7)*. Careful monitoring of conditions is mandated due to aromatization of ketone **35** under carbonylation conditions.

C-Glycosyl ester **31** (also **32** and **33**) can be employed for the synthesis of a variety of heteroaromatic analogs as indicated in Scheme 6. Esters have been transformed into isoxazole, imidazole, thiazole, and tetrazine derivatives employing established protocols *(24)*. Since both anomers of the glucopyranosyl system are available from the manganese methodology, it should be possible to prepare an array of C-glycosyl analogs for biological evaluation.

Scheme 4

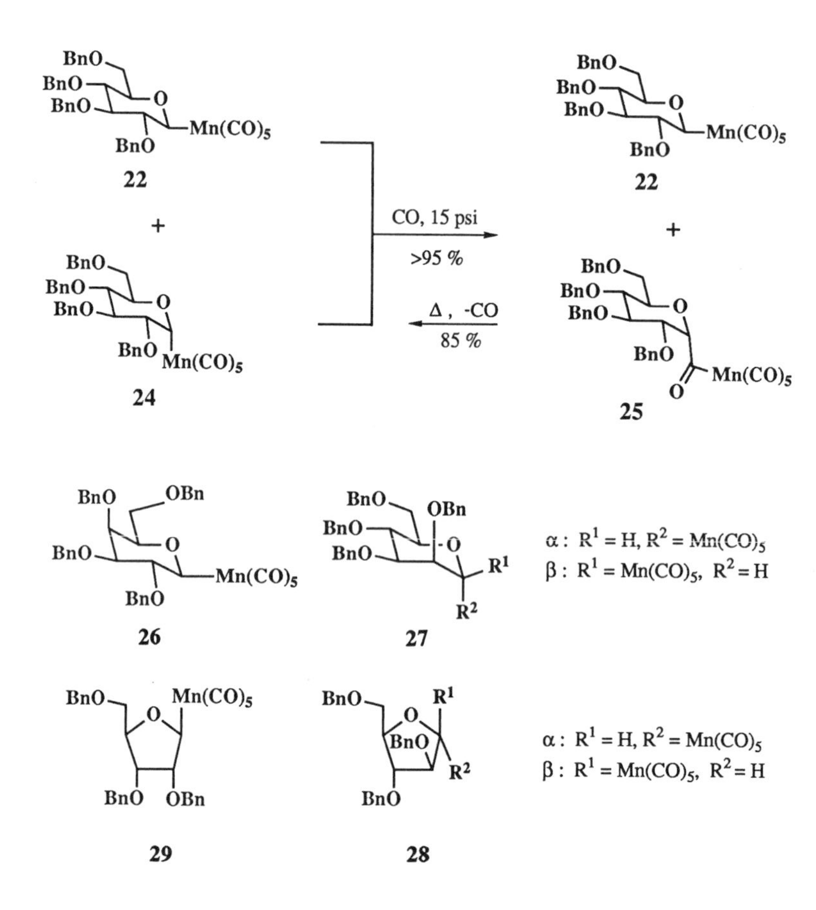
22
24
CO, 15 psi
>95 %
Δ , -CO
85 %
22
25
26
27
α : R1 = H, R2 = Mn(CO)5
β : R1 = Mn(CO)5, R2 = H
29
28
α : R1 = H, R2 = Mn(CO)5
β : R1 = Mn(CO)5, R2 = H

Scheme 5

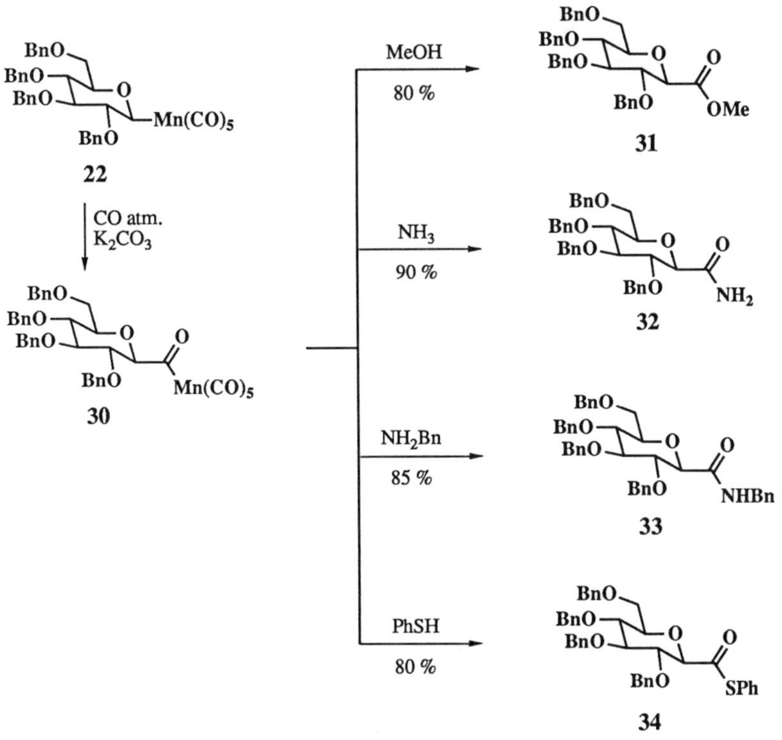

BnO BnO BnO O $Mn(CO)_5$ BnO

22

CO atm.
K_2CO_3

BnO BnO BnO O O BnO $Mn(CO)_5$

30

MeOH
80 %

BnO BnO BnO O O BnO OMe

31

NH_3
90 %

BnO BnO BnO O O BnO NH_2

32

NH_2Bn
85 %

BnO BnO BnO O O BnO NHBn

33

PhSH
80 %

BnO BnO BnO O O BnO SPh

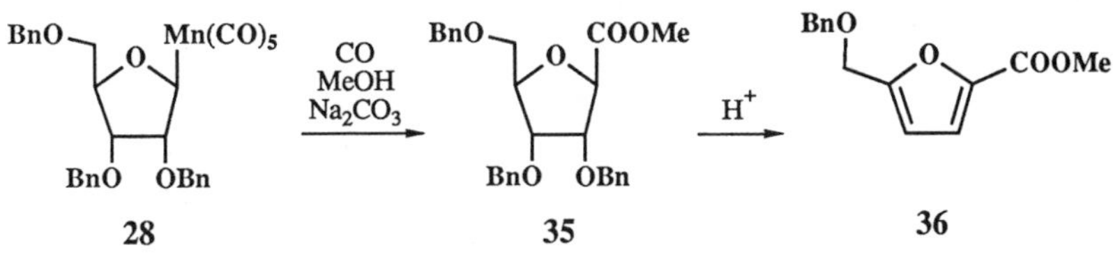

34

BnO O $Mn(CO)_5$ BnO OBn

28

CO
MeOH
Na_2CO_3

BnO O COOMe BnO OBn

35

H^+

BnO O COOMe

36

Scheme 6

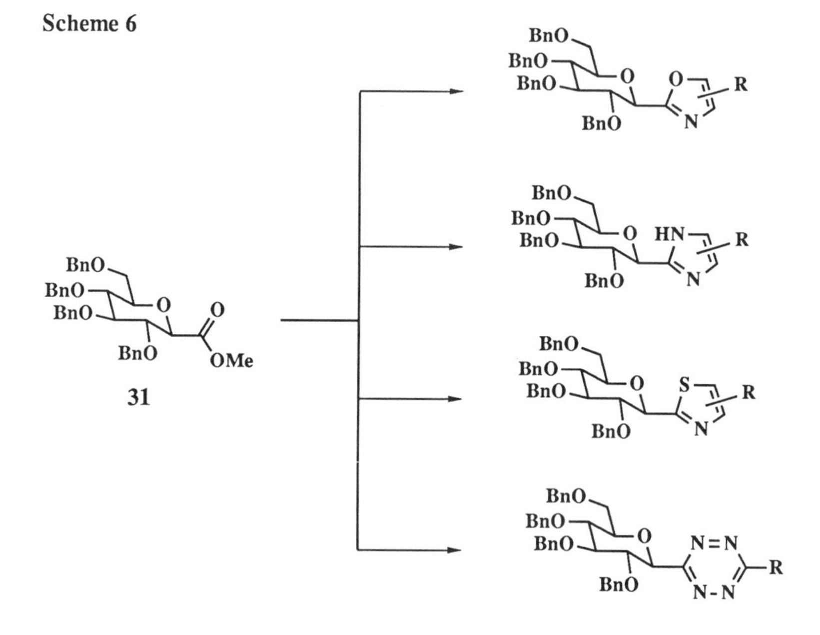

In the carbonylation of glycosylmanganese pentacarbonyls, a single carbon-carbon bond is formed. Sequential insertion, on the other hand, produces at least two new carbon-carbon bonds in the manganacycle products and it was gratifying to determine that in most instances the glycosyl complexes underwent sequential insertion efficiently. Scheme 7 summarizes a represenative array of examples with glucopyranosyl complex **22.** As indicated, glucosylmanganese complex **22** undergoes sequential insertion with electron-deficient alkenes and terminal alkynes in analogy with the simple alkylmanganese complexes. Sequential insertion of methyl acrylate with **22** at 5 kbar *(25)* yielded a 1:1 mixture of diastereomeric complexes **37**. We had hoped that the stereogenic center (the anomeric center) undergoing migratory insertion would induce stereoselectivity in the sequential insertion process, but this was not observed. However, when methyl crotonate was employed as the sequential insertion partner with complex **22**, a 10:1 mixture of diastereomers **38** was produced. The relative stereochemistry of the major diastereomer has not yet been determined. The excellent diastereoselectivity observed in this reaction suggests it may be possible to induce remote stereogenic centers in other sequential insertion processes.

Sequential insertion of complex **22** with phenyl vinyl sulfone at 5 kbar provided a 1:1 mixture of diastereomeric manganacycles **39**. These adducts were not fully characterized due to their instability. However, the products of demetalation of this adduct were stable and their characterization will be discussed below *(20)*.

Phenyl acetylene underwent regioselective insertion with complex **22** at atmospheric pressure to afford unsaturated manganacycle **40**. This result was analogous to the situation in simple alkylmanganese complexes where alkynes are more reactive toward sequential insertion and do not require high pressures to obtain good yields of maganacycles (Scheme 1) *(10, 12)*.

As anticipated, demetalation of manganacycle **37** could be accomplished photochemically in the presence of oxygen and water to produce enone **41** *(17)*. Alternatively, if the demetalation was performed *in the absence of oxygen*, then ketone **42** was the sole product of demetalation (Scheme 8). A discussion of the mechanism of this demetalation will be published elsewhere*(17)*.

Treatment of enone **42** with hydrazine (or its derivatives) gave dihydropyridazine **43** in good yield. Oxidation to pyridazine **44** has proven to be problematic in this instance. Enone **41**, on the other hand, reacts with hydrazine to afford dihydropyrazole **45** as a 1:1 mixture of diastereomers *(20)*. (See Scheme 9.)

Crotonate-derived manganacycle **38** undergoes an analogous series of transformations to those discussed in Scheme 8 which result in formation of methyl analogs of **41-44**.

Manganacycle **39,** which results from insertion of phenyl vinyl sulfone with β-complex **22** (Scheme 10), was unstable and could not be fully charaterized (*vide supra*). Photolysis of the crude manganacycle according to standard protocols provides keto-sulfone **46**. Sulfone **46** is prone to loss of phenylsulfinic acid to give enone **47***(20)*.

When manganacycle **39** is allowed to remain in ether solution in the dark at atmospheric pressure, a single diastereomer of butenolide **49**(configuration unknown) is produced in excellent yield*(20)*. Presumably, the butenolide results from formation of manganacycle **48** by loss of phenyl sulfinic acid, protonation of **48**, migratory insertion of an additional molecule of CO producing ketene **51**, enolization, and ring formation. We have previously demonstrated that this mechanism is operative in reactions of acetylene derivatives*(15)*.

Butenolide **49** can be treated with TBSOTf which results in pyran cleavage and formation of butenolide **52** as a mixture of geometric isomers (Scheme 11). Alternatively, reduction of the lactone carbonyl with DIBAL and dehydration affords C-glycosyl furan **53***(20)*.

To date, we have focused upon transforming β-glucopyanosyl complex **22** into a variety of C-glycosyl derivatives which will serve as one component in the Diels-Alder reaction for preparation of C-arylated glycosyl derivatives. These studies

Scheme 7

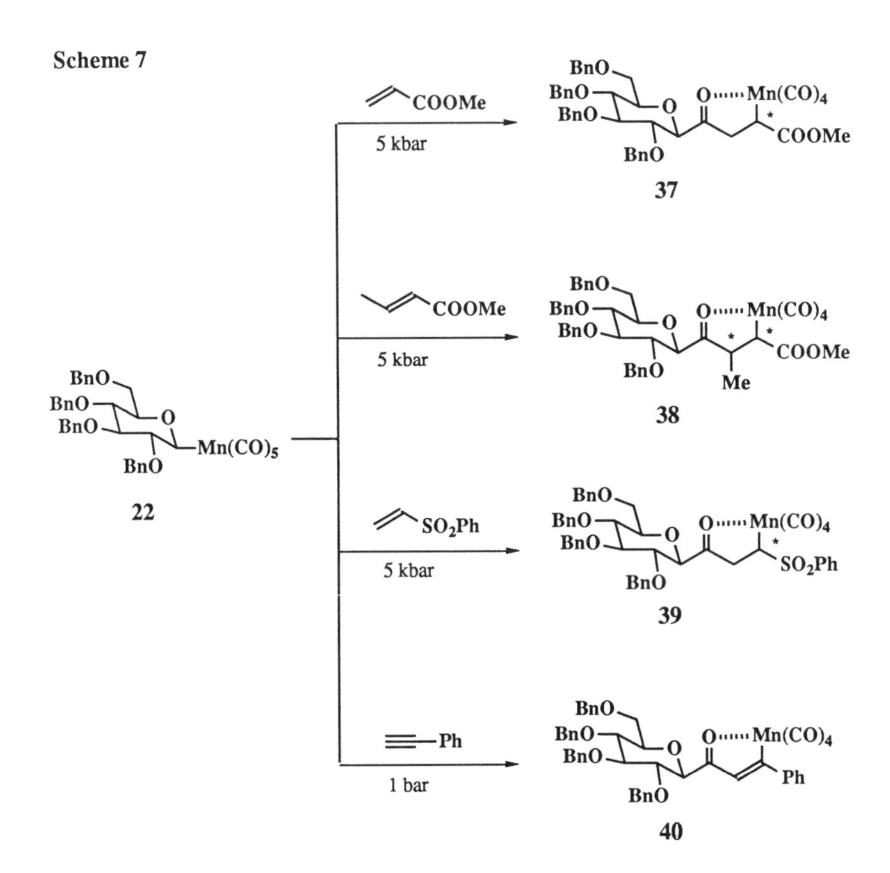
COOMe
5 kbar
COOMe
5 kbar
SO2Ph
5 kbar
Ph
1 bar
BnO
Mn(CO)5
22
O
Mn(CO)4
COOMe
37
Me
38
SO2Ph
39
Ph
40

Scheme 8

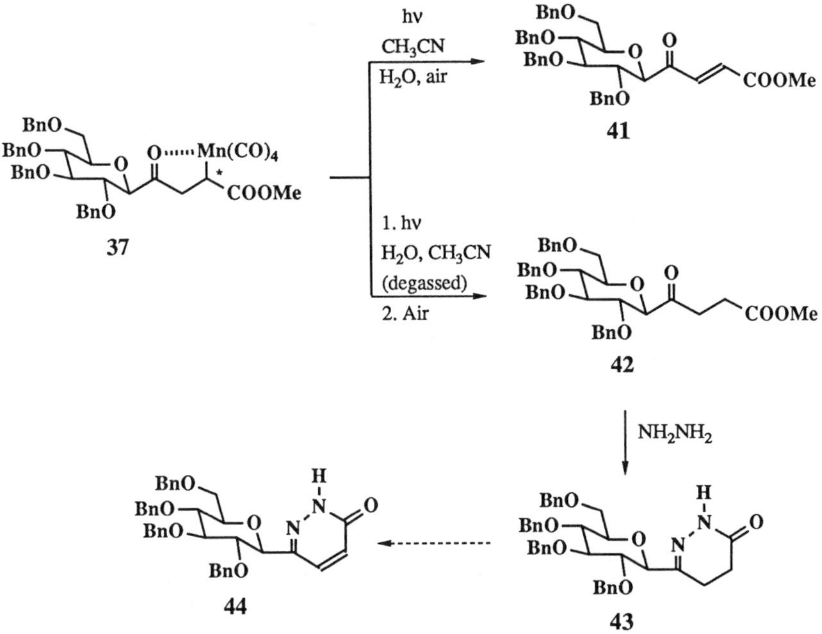

37

hν, CH_3CN, H_2O, air → **41**

1. hν, H_2O, CH_3CN (degassed); 2. Air → **42**

42 → NH_2NH_2 → **43**

43 - - - → **44**

Scheme 9

41 → NH_2NH_2, 86 % → **45**

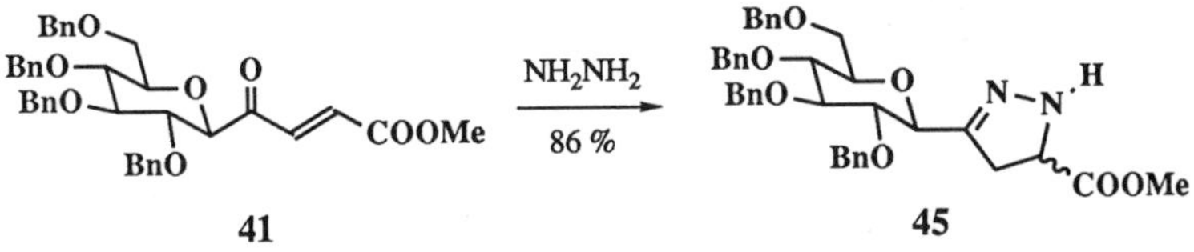

Model for Pyrazofurin

Scheme 10

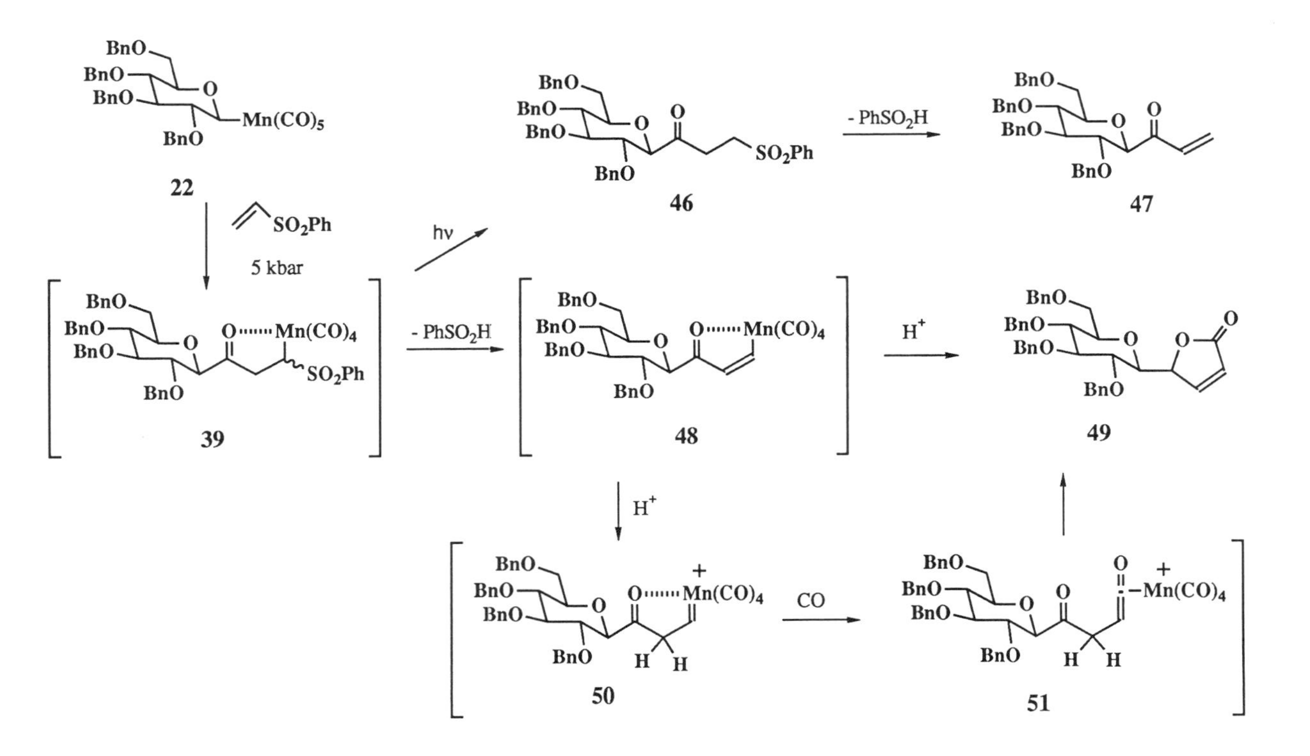

BnO
BnO
BnO
BnO
Mn(CO)5
22
SO2Ph
5 kbar
O
Mn(CO)4
SO2Ph
39
hv
- PhSO2H
46
- PhSO2H
47
48
H+
49
H+
50
CO
51

Scheme 11

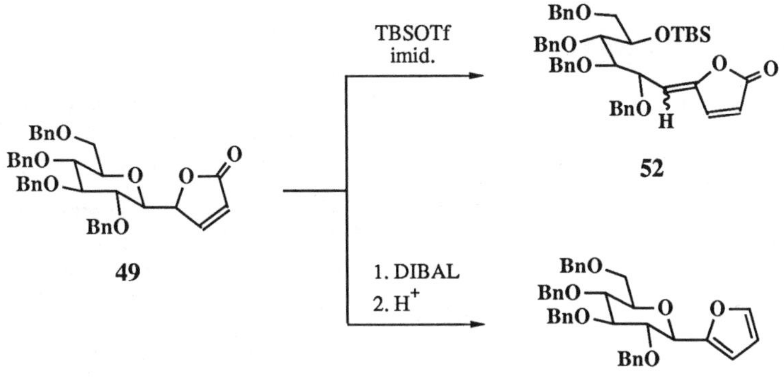

TBSOTf
imid.

BnO
BnO
BnO
BnO
O
O
O
49

BnO
BnO
BnO
OTBS
O
O
BnO
H
52

1. DIBAL
2. H^+

BnO
BnO
BnO
O
O
BnO
53

Scheme 12

BnO
BnO
BnO
O
$-Mn(CO)_5$
BnO
22

BnO
BnO
BnO
O
O
N
R
BnO

BnO
BnO
BnO
O
N=N
N-N
R
BnO

BnO
BnO
BnO
O
O
BnO

BnO
BnO
BnO
O
O
BnO

BnO
BnO
BnO
O
R
BnO

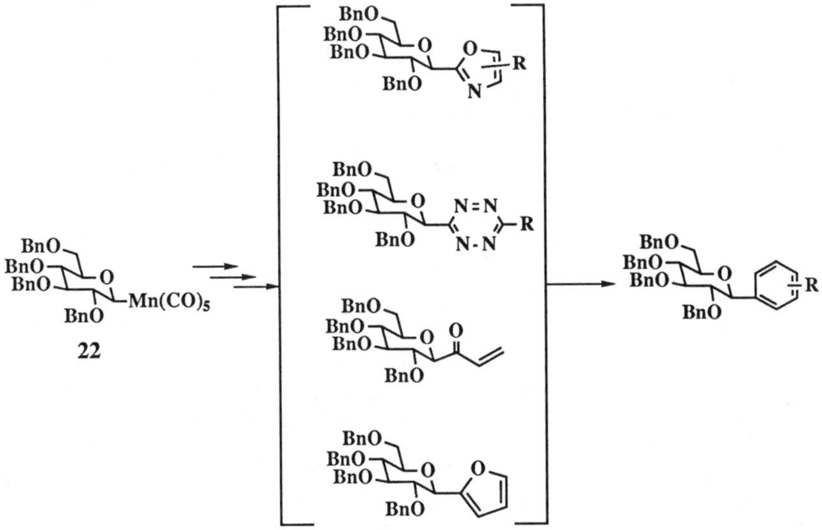

have resulted in the stereo- and regioselective synthesis of a variety of compounds which we anticipate can be applied to the synthesis of C-glycosyl antibiotics such as gilvocarcin and vineomycinone in the foreseeable future. (See Scheme 12.)

Acknowledgments

We thank the National Institutes of Health for generous financial support of this program. We also acknowledge financial support from Tektronix, Inc. and Lederle Laboratories.

References and Notes

1. Rosenthal, A. "Advances in Carbohydrate Chemistry and Biochemistry" Academic Press, New York, 1968, vol. 23, pp. 59-114. Suhadolnik, R. J. "Nucleosides as Biological Probes" Wiley Interscience, New York, 1979.

2. Goodchild, J. "The Biochemistry of Nucleoside Antibiotics" in "Topics in Antiobiotic Chemistry" Sammes, P. G., Ed. Halsted Press, New York, 1982, vol. 6, pp. 99-228. Hannessian, S.; Pernet, A. G. *Adv. Carbohydrate Chem. Biochem.* **1976**, *33*, 111. Buchanan, J. G. "The Nucleoside Antibiotics" in *Progress in the Chemistry of Natural Products*, **1983**, *44*, 243. Buchanan, J.G.; Wightman, R. H. "The Chemistry of Nucleoside Antibiotics" in "Topics in Antibiotic Chemistry" Sammes, P. G., Ed. Halsted Press, New York, 1982, vol. 6, 229-339.

3. Ellis, G. P.; West, G. P. "*Progress in Medicinal Chemistry*" North-Holland, Amsterdam. London, 1973, vol. 9, pp. 1-63.

4. Matsumoto, T.; Kakigi, H.; Suzuki, K. *Tetrahedron Lett.* **1991**, *32*, 4337. Matsumoto, T.; Katsuki, M.; Jona, H.; Suzuki, K. *Ibid.* **1989**, *30*, 6185. Danishefsky, S. J.; Uang, B. J.; Quallich, G. *J. Am. Chem. Soc.* **1985**, *107*, 1285. Tius, M. A.; Galeno, J. G.; Zaidi, J. H. *Tetrahedron Lett.* **1989**, *29*, 6909. Krohn, K.; Baltus, W. *Tetrahedron* **1988**, *44*, 49.

5. Jung, M. E.; Jung, Y. H. *Tetrahedron Lett.* **1888**, *29*, 2517 and references cited therein.

6. Farkas, J.; Flegelova, A.; Sorm, F. *Tetrahedron Lett.* **1972**, 2279. DeBernardo, S.; Weigele, M. *J. Org. Chem.* **1976**, *41*, 287. Buchanan, J. G.; Stobie, A.; Wightman, R. H. *J. Chem. Soc., Chem. Commun.* **1980**, 916. Katagiri, N.; Takashimi, K.; Haneda, T.; Kato, T. *J. Chem. Soc., Perkin Trans. 1* **1984**, 553.

7. Srivastava, P. C.; Pickering, M. V.; Allen, L. B.; Streeter, D. G.; Campbell, M. T.; Witkowski, J. T.; Sidwell, R. W.; Robins, R. K. *J. Med Chem.* **1977**, *20*, 256. Srivastava, P. C.; Robins, R. K. *J. Med . Chem.* **1983**, *26*, 445.

8. DeShong, P.; Slough, G. A.; Elango, V.;Trainor, G. *J. Am. Chem. Soc.* **1985**, *107*, 7788.

9. DeShong, P.; Slough, G.A.; Rheingold, A. L.*Tetrahedron Lett.* **1987**, *28*, 2229.

10. DeShong, P.; Sidler, D. R.; Slough, G. A.*Tetrahedron Lett.* **1987**, *28*, 2233.

11. DeShong, P.; Slough, G. A.; Elango, V. *Carbohyd. Res.* **1987**, *171*, 342.

12. DeShong, P.; Sidler, D. R.; Rybczynski, P. J.; Slough, G. A.; Rheingold, A. L. *J. Am. Chem. Soc.* **1988**, *110*, 2575.

13. DeShong, P.; Sidler, D. R. *J. Org. Chem.* **1988**, *53*, 4892.

14. DeShong, P.; Slough, G.A.; Sidler, D.R.; Rybczynski, P.J.; Philipsborn, W. v.; Kunz, R. W.; Bursten, B. E.; Clayton, T.W., Jr. *Organometallics* **1989**, *8*, 1381.

15. DeShong, P.; Sidler, D. R.; Rybczynski, P. J.; Ogilvie, A. A.; Philipsborn, W. v. *J. Org. Chem.* **1989**, *54*, 5432.

16. DeShong, P.; Rybczynski, P. J. *J. Org. Chem.* **1991**, *56*, 3207.

17. L. J. Smith, P. DeShong, unpublished results.

18. Trainor, G. L.; Smart, B. E. *J. Org. Chem.* **1983**, *48*, 2447.

19. Bock, P. L.; Boschetto, D. J.; Rasmussen, J. R.; Demers, J. P.; Whitesides, G. M. *J. Am. Chem. Soc.* **1974**, *96*, 2814.

20. T. A. Lessen, P. DeShong, unpublished results.

21. Lemieux, R. U.; Haymi, J. I. *Can. J. Chem.* **1965**, *43*, 2162. Lemieux, R. U.; Hendriks, K. B.; Stick, R. V.; James, K. *J. Am. Chem. Soc.* **1975**, *97*, 4056.

22. G. B. Anderson, P. DeShong, unpublished results.

23. T. X. Le, P. DeShong, unpublished results.

24. Metzger, J. V. "*Thiazoles and their Benzo Derivatives*" in "*Comprehensive Heterocyclic Chemistry*" Pott, K. T., Ed. Pergamon Press, NewYork, vol. 6, 1984.

25. DeShong, P.; Dicken, C. M.; Perez, J. J.; Shoff, R. M. *Org. Prep. Proce. Int.* **1982**, *14*, 369.

RECEIVED December 2, 1991

Chapter 8

Inter- and Intramolecular Diels–Alder Reactions of Sugar Derivatives

Pál Herczegh[1], Martina Zsély[1], László Szilágyi[2], István Bajza[2], Árpád Kovács[1], Gyula Batta[1], and Rezső Bognár[1,3]

[1]Research Group for Antibiotics of the Hungarian Academy of Sciences, H–4010 Debrecen, Hungary
[2]Department of Organic Chemistry, L. Kossuth University, H–4010 Debrecen, Hungary

Chiral thiopyran derivatives have been obtained in Diels-Alder reactions from monosaccharide O-thioformates and dithiooxalates. Intramolecular Diels-Alder cyclization of 1,6,8-nonatrienes and 1,7,9-decatrienes prepared from pentoses or D-glucose afforded multichiral hexahydroindene and octahydronaphthalene derivatives with very high diastereoselectivity.

Hetero Diels-Alder Reactions of Sugar-based Thiodienophiles

Because of their interesting biological effects, saccharides containing sulfur in the pyranose or furanose ring have been the subjects of a great deal of synthetic effort in carbohydrate chemistry (*1,2*) so far. Generally, the ring sulfur atom is introduced by means of nucleophilic substitution reaction of a properly protected sulfonate or halide derivative with thio nucleophiles. Another possibility is the direct construction of the thiopyranose ring through cycloaddition of a thiodienophile onto a diene. The first examples of the latter approach are represented by papers from Vyas and Hay (*3-6*). These authors applied methyl cyanodithioformate as dienophile for the synthesis of 6-thiodeoxyulopyranosidonic nitriles.

A few years ago we set out to investigate this method with the aim of generalization for the synthesis of other thiosugars. As a first attempt we have introduced new thiocarbonyl derivatives, the pentathiomonoorthooxalates **1** (*7*). These molecules contain a masked carboxylic function permitting carbanionic tranformations in a later stage of the synthesis without affecting the easily removable trithioorthoester grouping. **1a,b,c** have been obtained from trithioorthoformates **2** by deprotonation, reaction with carbon disulfide and subsequent methylation of the dithiocarboxylate as shown in Figure 1.

[3]Deceased

0097–6156/92/0494–0112$06.00/0

From the reaction of **1c** a crystalline side product could be isolated whose structure was established as **1c'** using selective INEPT measurements. The formation of **1c'** can be attributed to a concurrent deprotonation of a benzylic methylene or a rearrangement of the intermediate orthoformate anion to give **1c''**. Existence of **1c''** can be explained by steric crowding within the tribenzyl trithioorthoformate anion (Figure 2.). Attempted Diels-Alder reactions of **1a,b,c** with butadiene have, unfortunately, yielded the unsaturated, open-chain compounds **4a,b,c** instead of the expected thiopyran derivatives **3a,b,c.** We supposed that **3a,b,c** have been formed initially but due to the high steric compression of the trithiocarboxylate, an alkylthio group migration had taken place with concomitant ring opening of the sulfonium salt intermediate (Figure 3.).

The regio- and diastereoselectivity of reactions involving reactive groupings are often enhanced when such groups are attached to chiral carriers. For instance, disaccharides have been synthesized by David et al. using cycloadditions of monosaccharide butadienyl ethers (*8-10*). We wished to study the scope and limitations of this principle for the construction of chiral thiopyran derivatives, including disaccharide models, containing sulfur in a thiopyranoside ring.

For starting, we have chosen dithiooxalates and O-thioformates as dienophiles (*11*). Both types of compounds have been used for Diels-Alder reactions before (*12, 13*) but, to our knowledge, no attempt has been made to apply them to chiral substrates. Sugar dithiooxalates **6** and **8** as well as thioformates **9** and **10** have been prepared by known methods (*14, 15*) (Figure 4.)

Diels-Alder reactions of **6** and **8** with butadiene and 2,3-dimethylbutadiene under thermal conditions led to the expected mixtures of diastereomers **11a,b-14a,b.** Reaction of **6** and **8** with 2-trimethylsilyloxybutadiene proceeded, however with full regioselectivity and afforded mixtures of **15a,b** and **16a,b** after desilylation. (Figure 5.) Due to the presence of the chiral carrier a moderate diastereoselection could be observed in some cases. The diastereomeric mixtures could be separated only in the case of **16a,b**. The configurations of C-2 in the thiopyran part of **11a,b-16a,b** could not be deduced from the NMR spectra. In order to prove the structures of **15a,b** and **16a,b** we have used Raney-Ni desulfurization to get the expected 5-ketohexanoate esters (Figure 6.).

Using somewhat more drastic conditions (160°) reactions of the dienes with **5** and **7** gave rise to the formation of **17a,b-22a,b**, respectively. The extent of the diastereoselectivities were similar to those observed for **11a,b-16a,b** (Figure 7.) All of the stereoisomers in the glucose series **17a,b, 19a,b, 21a,b** could be separated by chromatography. An attempt has been made for the preparation of

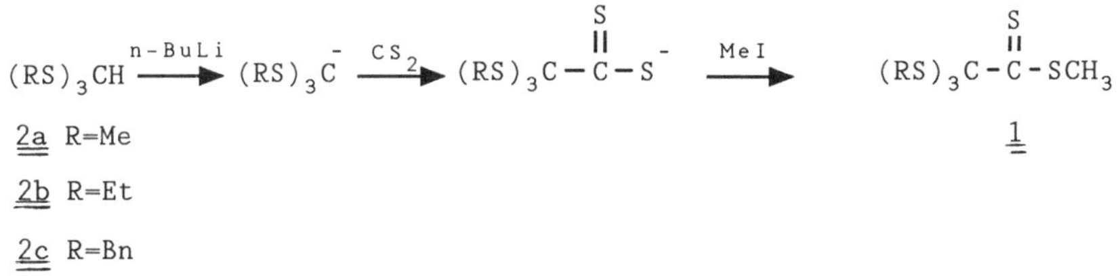
(RS)3CH
n-BuLi
(RS)3C
CS2
(RS)3C-C-S
S
MeI
(RS)3C-C-SCH3
S
1
2a R=Me
2b R=Et
2c R=Bn

Figure 1. Preparation of pentathiomonoorthooxalates.

C6H5-CH-SCH
SCH2C6H5
SCH2C6H5
C=S
SCH3
1C'
C6H5-CH-SCH
SCH2C6H5
SCH2C6H5
1C''

Figure 2. The side product **1c**'.

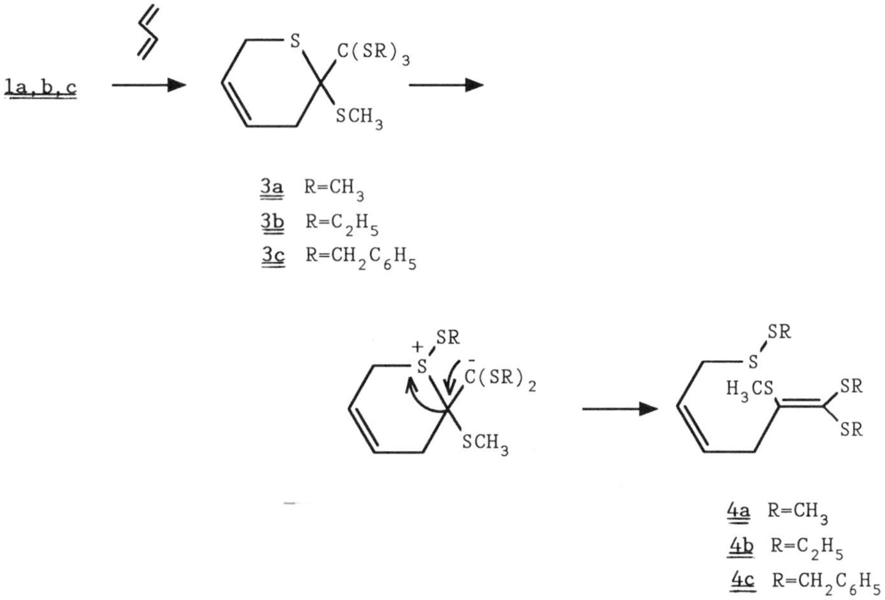
1a,b,c
S
C(SR)3
SCH3
3a R=CH3
3b R=C2H5
3c R=CH2C6H5
SR
S
C(SR)2
SCH3
SR
S
H3CS
SR
SR
4a R=CH3
4b R=C2H5
4c R=CH2C6H5

Figure 3. Attempted Diels-Alder reactions of pentathiomonoorthooxalates.

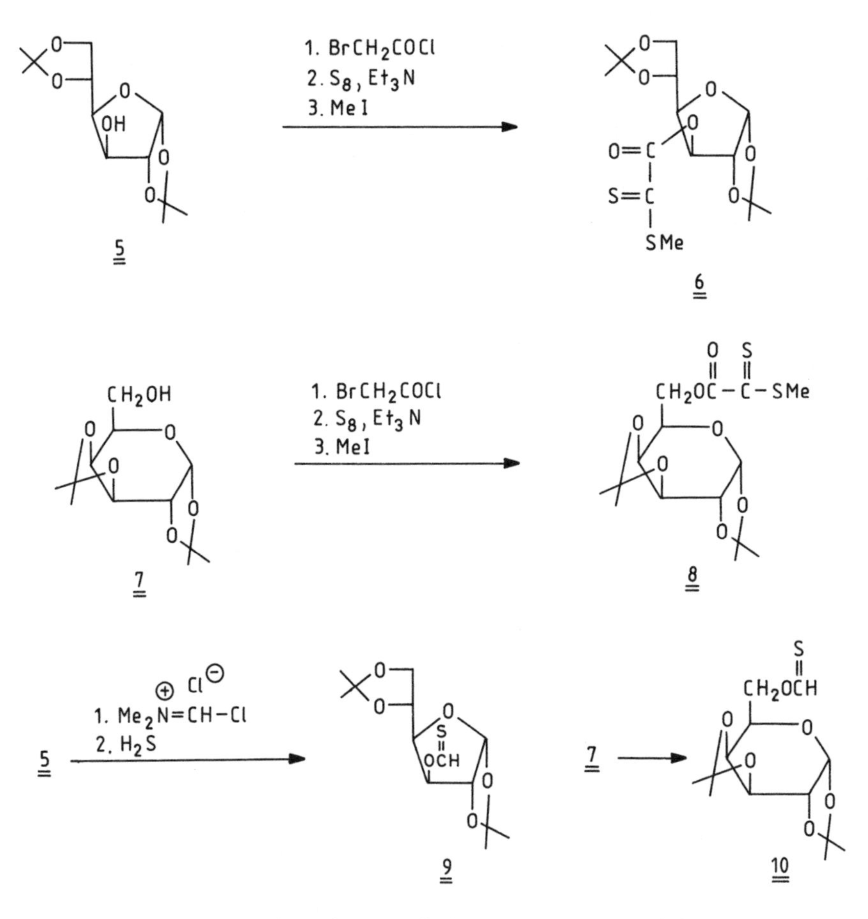

Figure 4. Preparation of sugar dithiooxalates and O-thioformates.

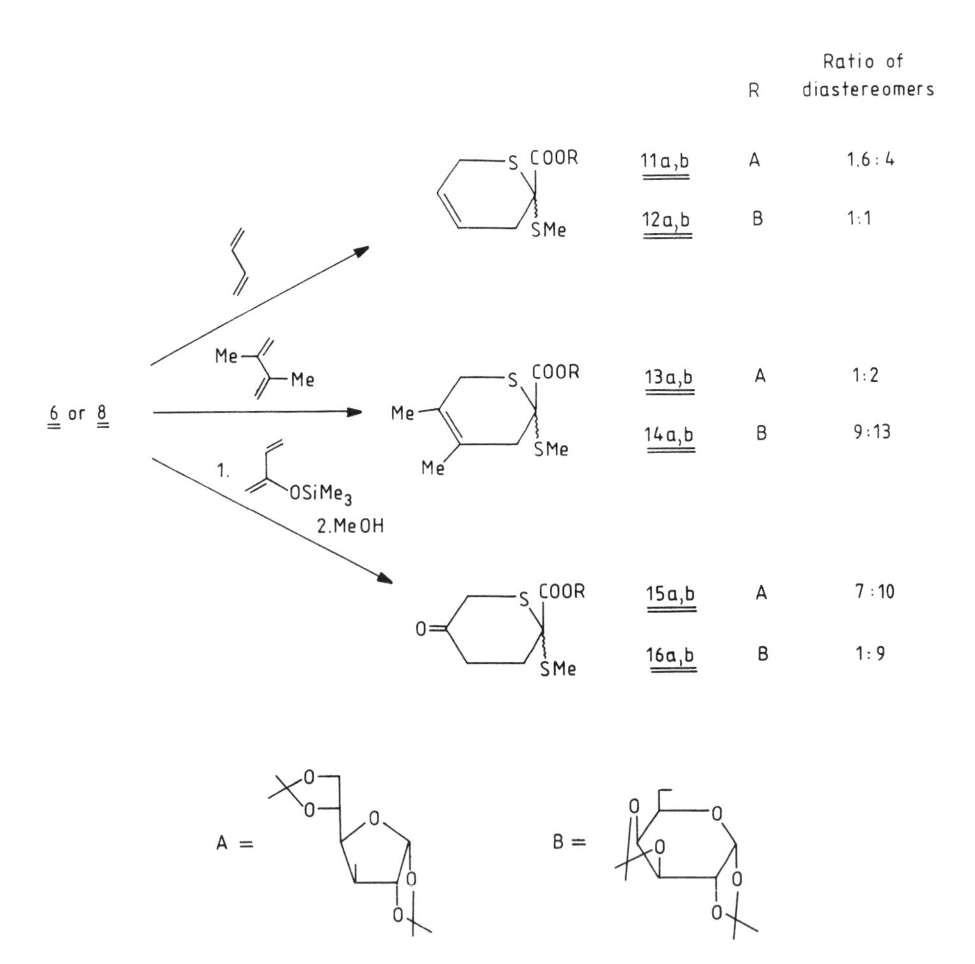

Figure 5. Hetero Diels-Alder reactions of sugar dithiooxalates.

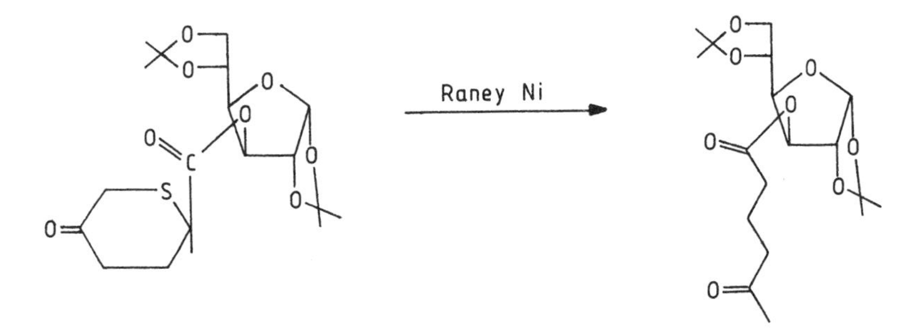

Figure 6. Proof of the regioselectivity of Diels-Alder reaction.

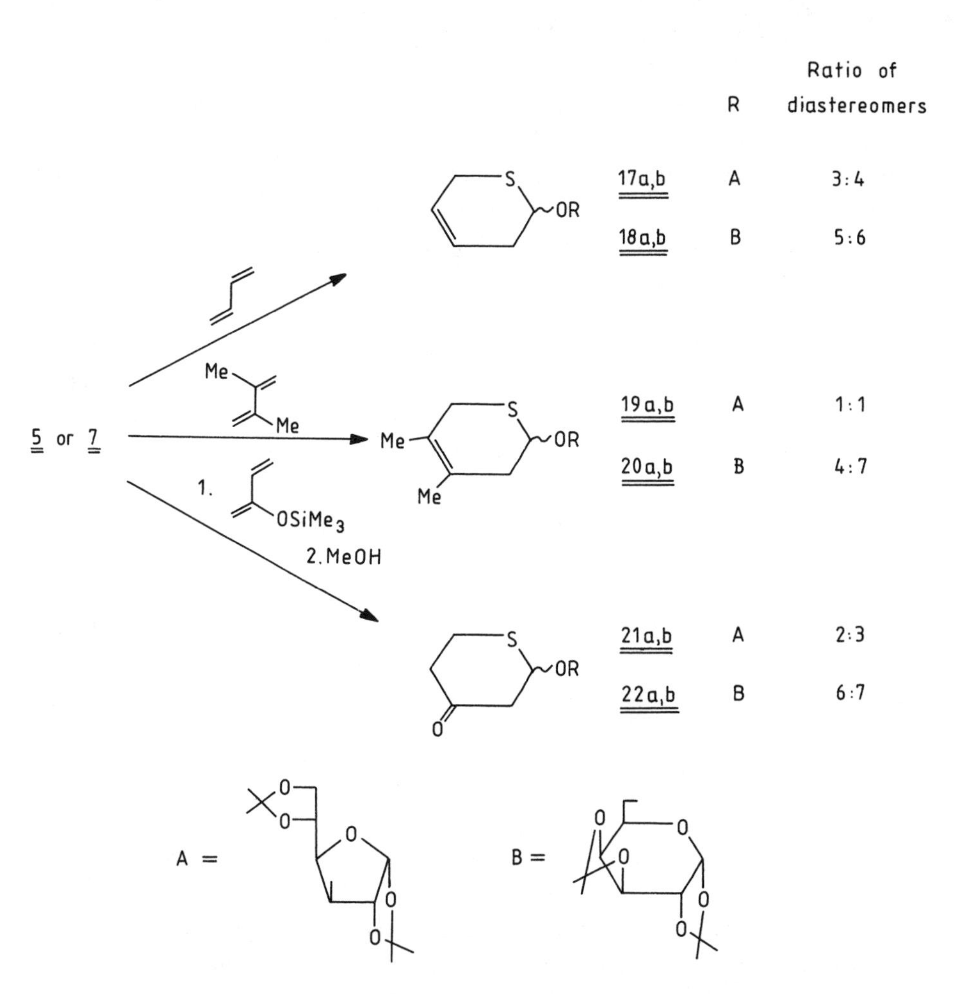

Figure 7. Hetero Diels-Alder reactions of sugar O-thioformates.

17a,b at room temperature under 2.5 kbar pressure. A promising (2:5), enhanced stereoselectivity was observed. The regioselectivities for the reactions of **5** and **7** with 2-trimethylsilyloxybutadiene were also complete as in the reaction of **6** and **8** but of opposite sense, leading to the thiopyran-4-ones **21a,b** and **22a,b**, respectively. The structure of the latter derivatives could be proved by NMR using selective INEPT measurements. Long-range couplings could be observed between H-2 and C-3 ($^2J_{C,H}$); H-2 and C-4($^3J_{C,H}$) and H-2 and C-6 ($^3J_{C,H}$), respectively, corroborating the location of the keto group in **21a,b** and **22a,b** as being in position 4 of the thiopyran ring. A possible explanation for the opposite regioselectivities of the thioformates and dithiooxalates with 2-trimethylsilyloxybutadiene is the following. Literature data (*16, 17*) suggest that in the Diels-Alder reactions between thiocarbonyl compounds and dienes having electron donating substituents in position 2, the preferred molecular orbitals are the LUMO of the thiodienophile (T-LUMO) and the HOMO of the diene (D-HOMO). We have performed *ab initio* calculations (Komáromi, I.; Herczegh, P., unpublished results) on model compounds **23** and **24**. Figure 8 shows the orbital coefficients at the T-LUMO-s. As it can be seen the ratios of the T-LUMO-s at the carbon and sulfur atoms of **23** and **24** are inverted. Thus, the opposite regioselectivity of the Diels-Alder reactions of **6,8** and **5,7** with 2-trimethylsilyloxybutadiene can be explained by the opposite polarities of their thiocarbonyl groups.

17a and **17b** could be separated by chromatography. *cis*-Hydroxylation by osmium(VIII)oxide gave diastereomeric mixtures of **25a,b** and **26a,b**, respectively. These derivatives are simple disaccharides bearing sulfur in the pentopyranose unit (Figure 9.).

We have thus established that thiocarbonyl dienophiles on chiral saccharide carriers can be used for the synthesis of optically active thiopyran derivatives including 5-thiopyranoses.

Diastereoselective intramolecular Diels-Alder reactions of sugar-derived trienes.

Intramolecular Diels-Alder (IMDA) reactions of 1,6,8-nonatrienes and 1,7,9-decatrienes have been used frequently for the construction of chiral ring systems of complex natural products (*18-20*). Although saccharide derivatives have been employed many times in intermolecular Diels-Alder reactions (*21-28*) their IMDA reaction appears not to have been studied so far. The diastereoselectivity of IMDA reactions can be influenced, among other factors, by the size of substituents, and by the chirality of carbons of the connecting chain. Our aim was to study the possibilities of the application of single monosaccharides as tether chains in IMDA reactions. The influence of the presence of three or four chiral

Figure 8. Orbital coefficients at T-LUMO.

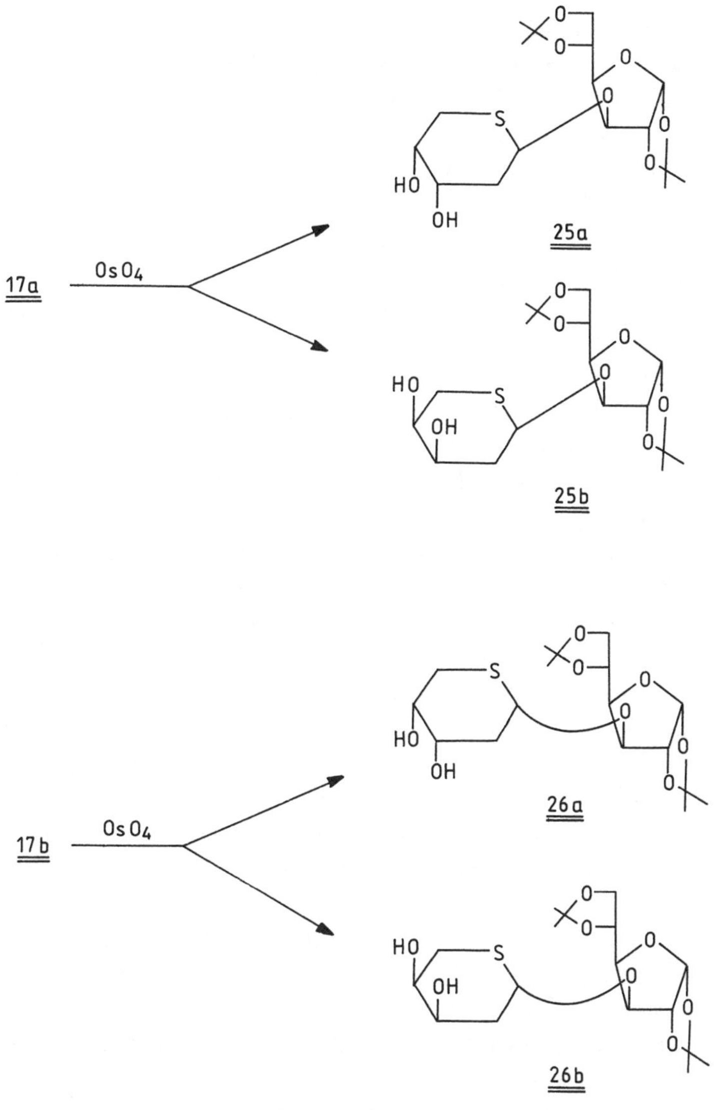

Figure 9. Synthesis of thiodisaccharides.

centers in the connecting chain on the stereochemical outcome of IMDA reactions has not been studied so far on enantiomerically pure substrates.

We have initially chosen three pentoses of the D-series as precursors for 1,6,8-nonatrienes. The D-xylose derivative **27** (*29*) has been mercaptalized (**28**), then the terminal hydroxyl of **28** oxidized to give the **29** dialdose which, in a Wittig reaction, gave rise to a mixture of E and Z dienes **30**. Demercaptalization (**31**) and another chain extension yielded an inseparable mixture (4:6) of E and Z nonatrienes **32** (Figure 10.). When **32** was subjected to heating at 160° it afforded, however, a single diastereomeric hexahydroindene **33** (Figure 11.). The structure of **33** can be deduced from the ^{1}H-NMR spectrum and from NOE experiments. As H-2, H-3a, H-7a shows strong reciprocal NOE-enhancements, the stereochemistry of the ring-fusion is clearly demonstrated. The proton-proton coupling constants are compatible with the envelope conformation of the cyclopentane part of the molecule. On the other hand, the magnitude of $J_{3a,7a}$ (8 Hz) speaks for the half-chair conformation of the cyclohexane ring.

The nature of the dienophilic part of the molecule did not have any influence on the stereochemistry of the IMDA reaction. **27** was reacted with methoxyamine, and the terminal hydroxyl of the obtained oxime was oxidized to give an aldehyde function (Figure 10.). Chain extension of the latter with allylidenetriphenylphosphorane led to the formation of a mixture of four diastereomers of **34**, i.e. E-syn, E-anti, Z-syn, Z-anti, respectively. A similar mixture has been obtained from **31** with methoxyamine. When the diastereomeric mixture of **34** was heated in toluene for 8 h the azabicyclo 4.3.0 nonene **35** diastereomer was formed exclusively. The conformation of this ring system is similar to that of **33** (*30,31*) (Figure 11.) as evidenced by $J_{1,6}$ (6.5 Hz).

Similarly high diastereoselectivity has been observed in the IMDA reactions of nonatrienes **37** and **41** derived from D-ribose and D-arabinose, respectively (Herczegh, P.; Kovács, Á., L. Szilágyi, unpublished results). Thus, the tri-O-benzyl ether **36** obtained from D-ribose has been transformed into the single Z-triene **37** by a sequence of oxidation of the terminal hydroxyl, chain extension with allylidenetriphenylphosphorane, demercaptalization and another Wittig reaction with methylenetriphenylphosphorane as depicted in Figure 12. Thermal IMDA reaction of **37** resulted in the exclusive formation of a *cis*-fused product **38**.

The triene **41** (9:1 mixture of Z and E isomers) has been prepared from the D-arabinose derivative **39** (*32*) in three steps. **40** has been obtained in a Wittig reaction. Oxidation of the terminal hydroxyl group and another chain extension led to **41** (Figure 13.). IMDA reaction of the latter at 120° gave the *cis*-fused product **42** together with trace amounts of an unidentified compound.

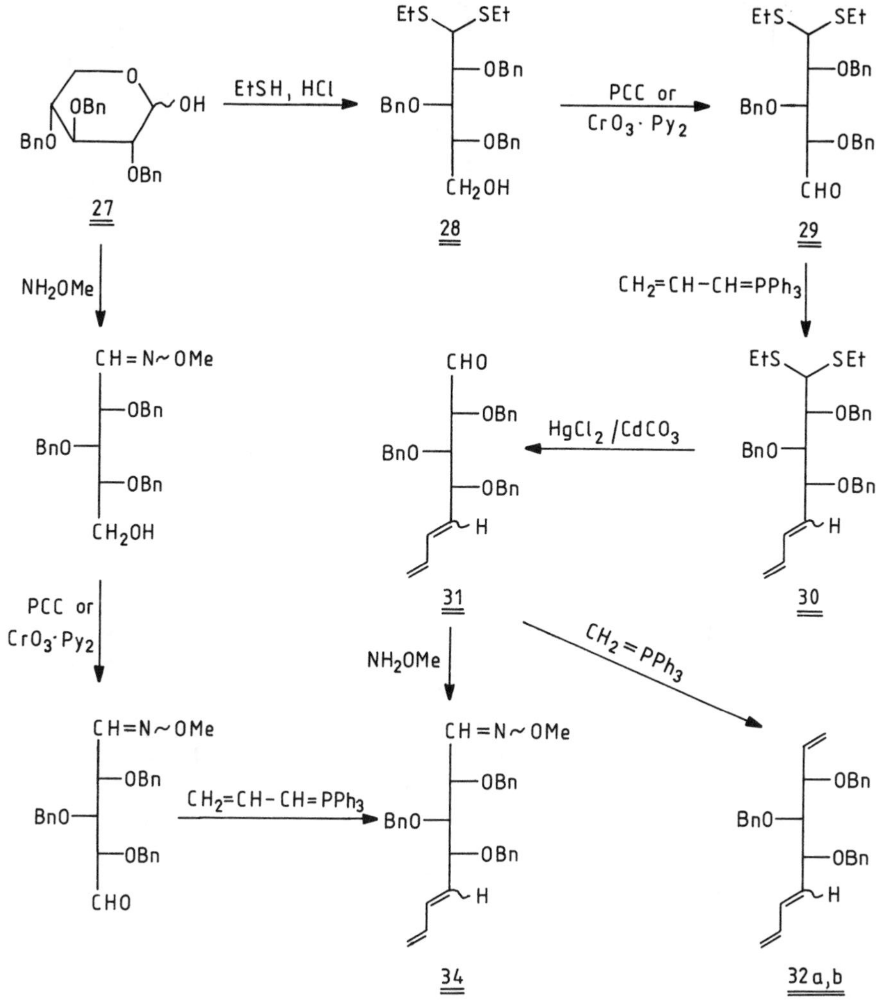

Figure 10. Preparation of tribenzyloxynonatrienes with *xylo* configuration.

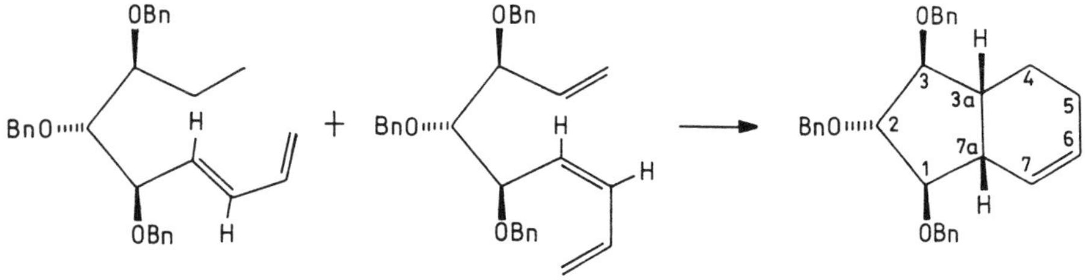

32a,b

33

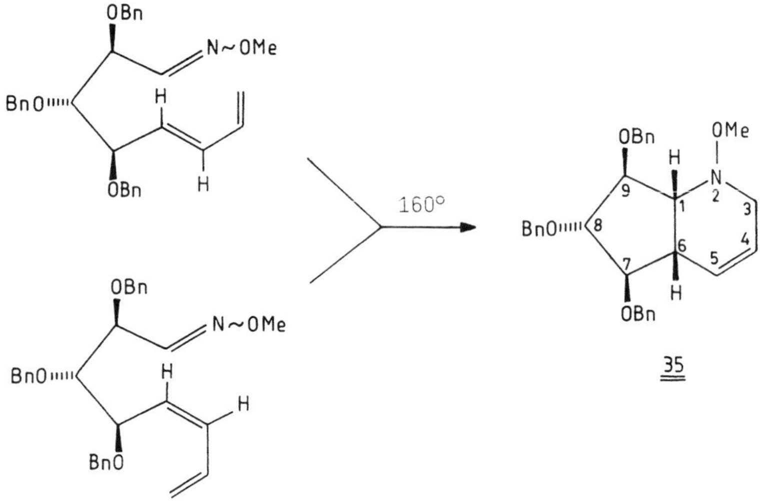

160°

34

35

Figure 11. IMDA reactions of the *xylo*-nonatriene.

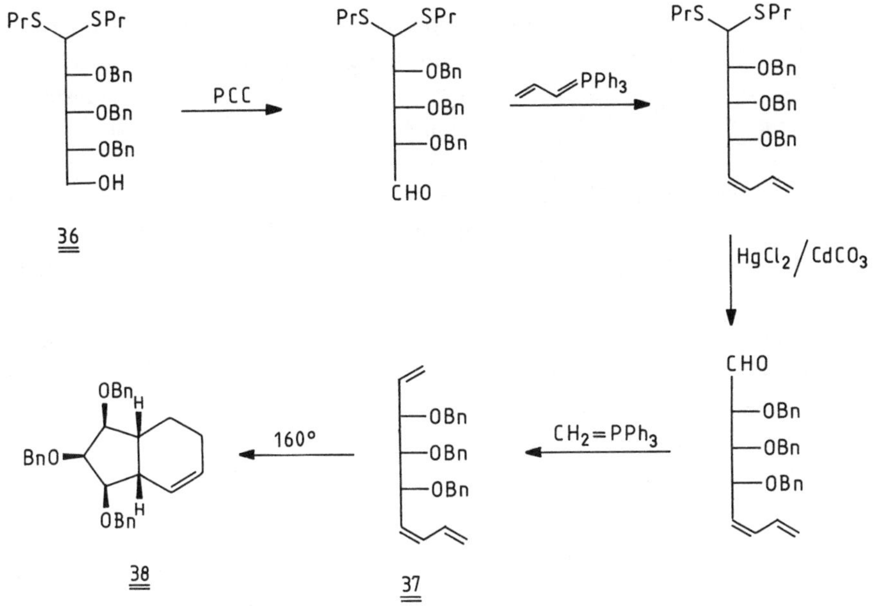

PrS SPr
OBn
OBn
OBn
OH
36
PCC
PrS SPr
OBn
OBn
OBn
CHO
PPh_3
PrS SPr
OBn
OBn
OBn
$HgCl_2/CdCO_3$
CHO
OBn
OBn
OBn
$CH_2{=}PPh_3$
OBn
OBn
OBn
37
160°
OBn H
BnO
OBn H
38

Figure 12. IMDA reaction of the *ribo*-nonatriene.

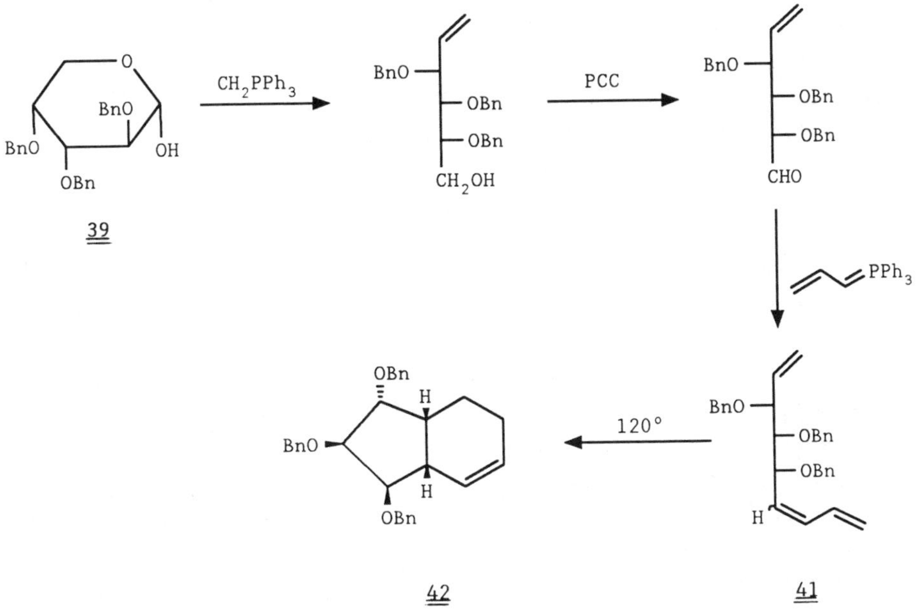

O
BnO
BnO
OH
OBn
39
CH_2PPh_3
BnO
OBn
OBn
CH_2OH
PCC
BnO
OBn
OBn
CHO
PPh_3
BnO
OBn
OBn
H
41
120°
OBn
H
BnO
H
OBn
42

Figure 13. IMDA reaction of the *arabino*-nonatriene.

With the results of the IMDA reactions of the tribenzyloxynonatrienes **32**, **37** and **41** in hand we attempted to explain the stereochemical outcome of those reactions by taking into consideration conformations leading to transition states.

For the 7E isomer of **32** conformation **A**, resulting the product **33** seems to be energetically the most favourable one because all of the substituents are oriented "equatorially" (Figure 14.). In **B** and **D** the eclipsed position of the 5-benzyloxy substituent and the adjacent diene moiety is a destabilizing factor. In **C** the "axial" 3-benzyloxy group strongly hinders the reactant parts from overlapping.

For the 7Z isomer conformer **A'** seems to be more favourable than **B'** because, in the latter, the 5-benzyloxy substituent lies between the diene and the dienophile preventing them from approaching each other.

It must be noted that conformations **A** and **A'** lead to the same product **33**. Therefore the stereochemistry of the product from the IMDA reaction of an E/Z mixture of **32** can be explained by this "degenerated" transition state. However, thermal Z/E isomerization cannot be excluded either.

The conformations **H** and **I** for the 7Z-triene **37**, depicted in Figure 15 represent a similar case: in the transition state **I** the 5-benzyloxy substituent completely hinders overlap with the reacting species, therefore only conformer **H** can lead to the bicyclic product **38**.

The formation of **42** in the IMDA reaction of **41** can be explained in a similar fashion. The two possible conformations **J** and **K** for the transition states of **41** can be seen in Figure 15. As for **B'** and **I** , hindering from the 5-benzyloxy group is a determining factor in the case of **K**. Thus, conformation **J** can be rationalized on the basis of the predominance of the formation of product **42**.

IMDA reactions of trienes derived from D-glucose

47a and **47b** have been prepared from **43** in four steps (Figure 16.). The first Wittig reaction led to a mixture of E and Z dienes **44a,b**. Debenzoylation, subsequent oxydation of the terminal hydroxyl group and another chain extension afforded again an E,Z mixture **47a,b** that could be separated by chromatography. Thermal IMDA reaction of the E isomer **47a** resulted in the formation of a sole product **49**. The same product has been obtained under similar conditions from the Z isomer (*33*).

An E,Z dodecatriene mixture **48a,b** has been prepared from **46a** by Wittig reaction with acetylmethylenetriphenylphosphorane. Thermal cyclization of **48a,b** gave rise to a single product **50** (Figure 17.). This stereoselectivity is remarkable since three new chiral centers have been formed simultaneously. The stereochemistries of **49** and **50** could be supported by NOE experiments and

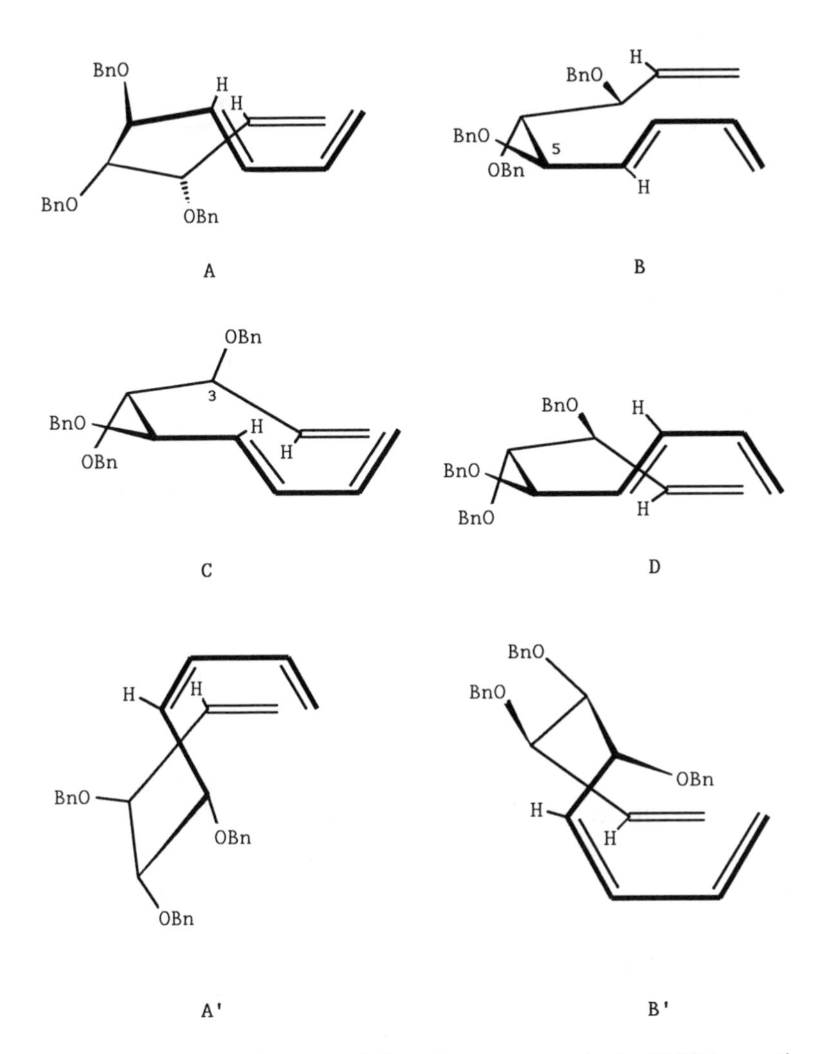

Figure 14. Possible conformers of the *xylo*-nonatriene in the IMDA reaction.

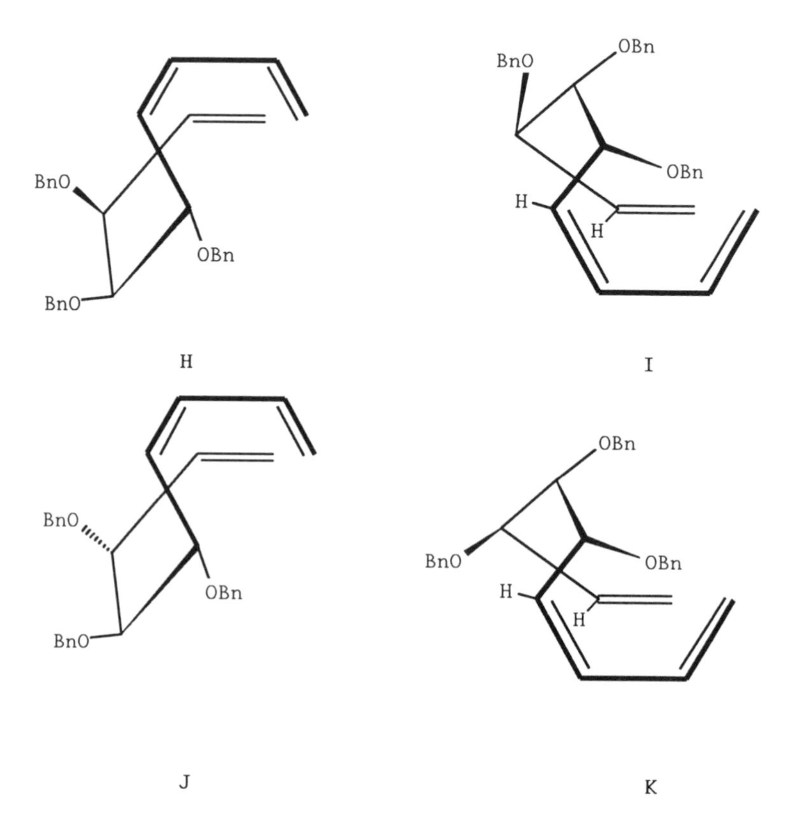

Figure 15. Possible conformers of the *ribo*- and *arabino*-nonatrienes in the IMDA reaction.

^{1}H-NMR coupling constants. For such types of octahydronaphthalenes the values of $J_{4a,8a}$ are usually less then 4 Hz in the *cis*-fused isomer and 11 Hz for the *trans*-fused ones. For **49** and **50** $J_{4a,8a}$ is 7-8 Hz. Because of the presence of dioxolane rings the cyclohexane ring presumably adopts a distorted boat conformation.

The sterochemical course of the IMDA reactions above can be explained by considering transition state conformations. For the E isomer **47a** transition state **L** seems to be energetically the most favourable one since all of the substituents are positioned "equatorially" and, unlike the situation seen in **M** and **O** there is no eclipsed crowding between the dienic part and the adjacent oxygen. In this way the reaction proceeds *via* a boat-like conformation which is preserved in the product. In **M** and **N** the "axially" oriented 2,3-dioxolane ring is a very destabilizing factor; **O**, on the other hand, suffers from an eclipsed crowding between the diene and O-6 as mentioned above (Figure 18.).

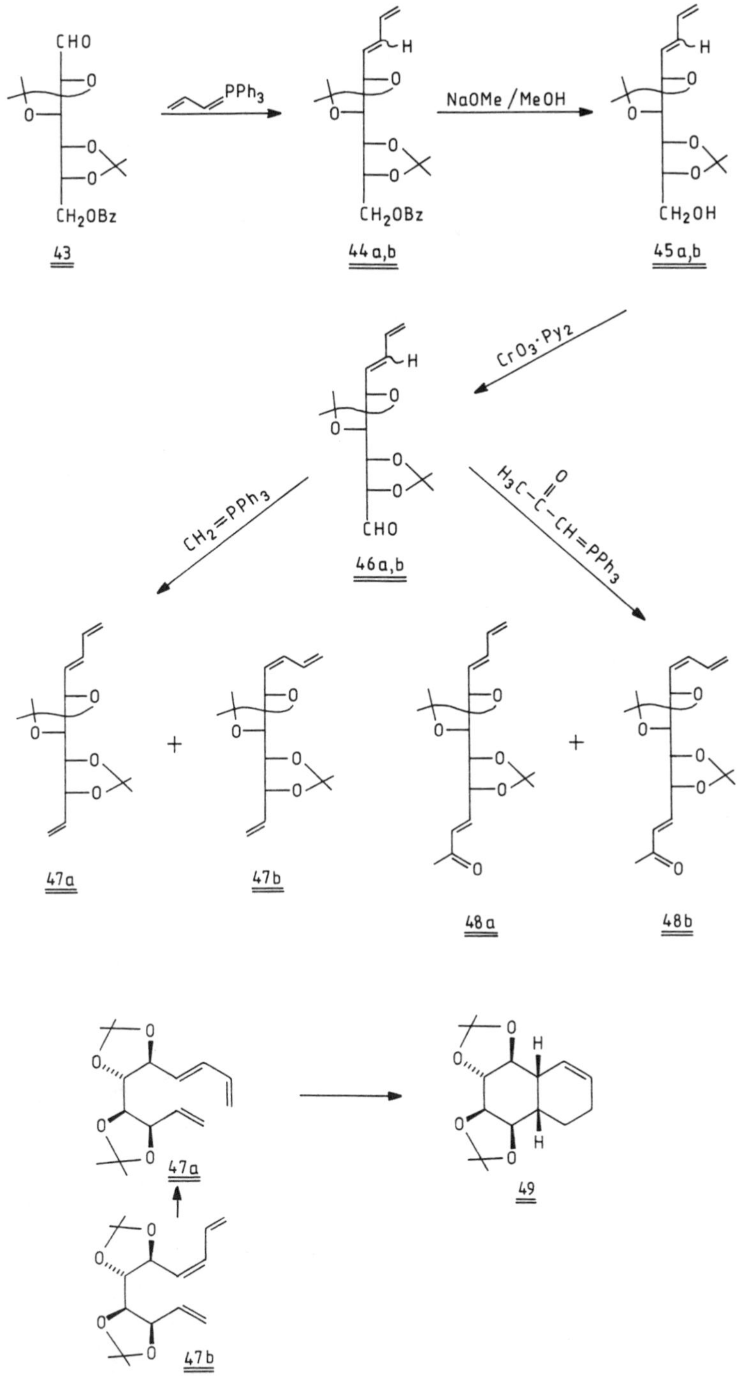

Figure 16. Preparation and IMDA reaction of decatrienes derived from D-glucose.

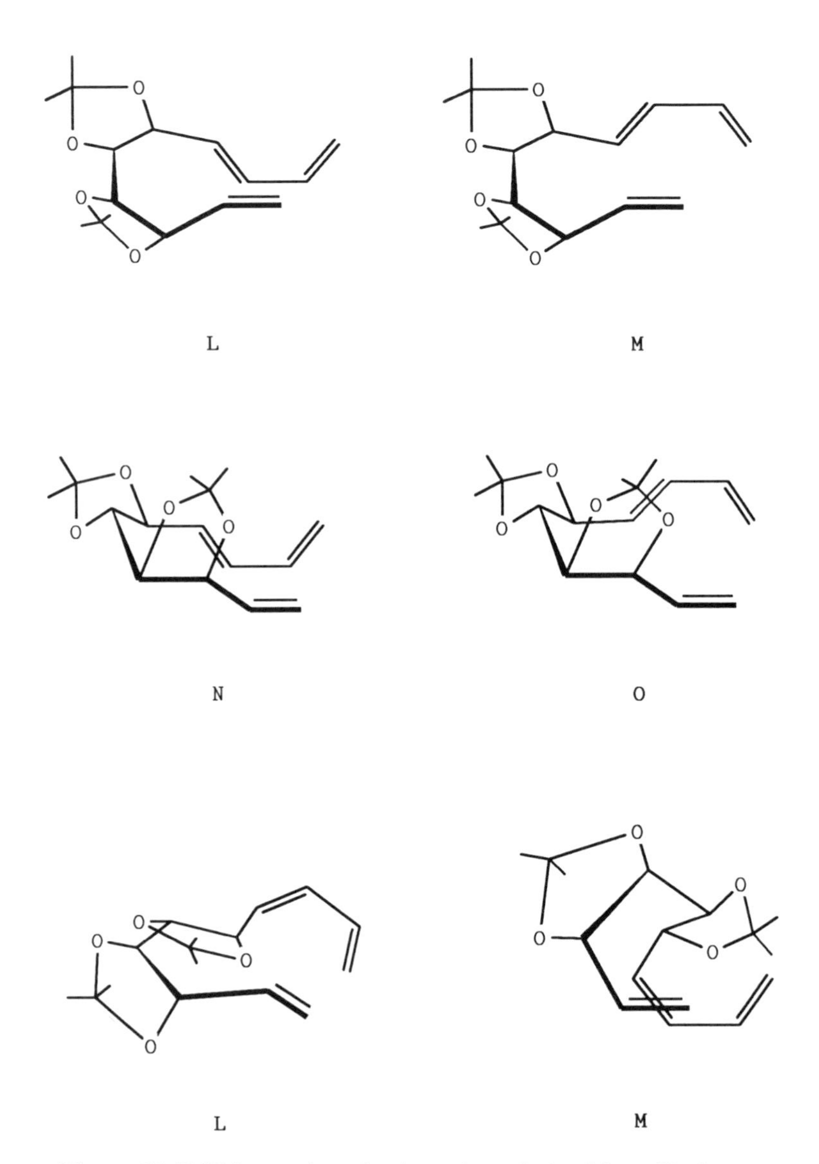

Figure 17. IMDA reaction of a decatriene derived from D-glucose.

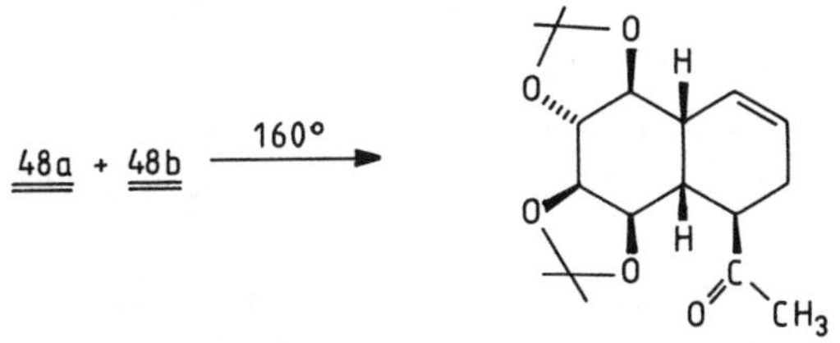

48a + 48b →(160°) 50

Figure 18. Possible conformations of the decatriene in the IMDA reaction.

For the Z isomer **47b** two conformers, **L'** and **M'** can be considered but, due to the presence of the two planar dioxolane rings they seem to be unsuitable for product formation. Owing to the fact that thermal cyclization of the Z diene still afforded **49** a thermal Z/E isomerization could be assumed. Indeed, GC-MS monitoring of the reaction of **47b** showed simultaneous formation of the E isomer **47a** and the product. To our knowledge this phenomenon has been observed in only one case (*34, 35*) thus far.

In summary it could be established that the stereochemical course of thermal IMDA reactions can be strongly influenced by substituents in the linking chain. Therefore, sugar-based trienes seem to be promising starting materials for the construction of multichiral hexahydroindene and octahydronaphthalene derivatives.

Literature Cited

1. Daniel, J.R.; Whistler, R.L.; Zingaro, R.A. *Phosphorus and Sulfur* **1979**,*7*, 31.
2. Paulsen, H. *Angew. Chem.* **1966,** *78,* 501.
3. Vyas, D.M.; Hay, G.W. *Canad. J. Chem.* **1971,** *49,* 3755.
4. Vyas, D.M.; Hay, G.W. *J. Chem. Soc., Perkin Trans. I.* **1975,** 180.
5. Vyas, D.M.; Hay, G.W. *Canad. J. Chem.* **1975,** *53,* 1362.
6. Vyas, D.M.; Hay, G.W. *Carbohydr. Res.* **1977,** *55,* 215.
7. Herczegh, P.; Zsély, M.; Szilágyi, L. *Tetrahedron* **1988,** *44,* 2063.
8. David, S.; Lubineau, A.; Vatele, J.M. *J. Chem. Soc., Chem. Commun.* **1975,** 701.
9. David, S.; Eustache, A.; Lubineau, A. *J. Chem. Soc., Perkin Trans. I.* **1979,** 1795.

10. David, S.; Lubineau, A. *J. Org. Chem.* **1979,** *44,* 4986.
11. Herczegh, P.; Zsély, M.; Szilágyi, L.; Bognár, R. *Heterocycles* **1989,** 887.
12. Hartke, K.; Quante, J.; Kampchen,T. *Liebigs Ann. Chem.* **1980,** 1482.
13. Vedejs, E.; Arnot, M.J.; Dolphin, J.M.; Eustache, J. *J. Org. Chem.* **1980,** *45,* 2601.
14. Barton, D.H.R.; McCombie, S.W. *J. Chem. Soc., Perkin Trans. I.* **1975,** 1575.
15. Thiel, W.; Viola, H.; Mayer, R. *Z. Chem.* **1977,** *17,* 366.
16. Rao, V.P; Chandrasekhar, J.; Ramamurthy, V. *J Chem. Soc., Perkin Trans. II.* **1988,** 647
17. Vedejs, E.; Perry, D.A.; Houk, K.N.; Rondan, N.G. *J. Am. Chem. Soc.* **1983,** *105,* 6999.
18. Taber, D.F. *Intramolecular Diels-Alder Reactions and Alder Ene Reactions;* Springer-Verlag, Berlin-Heidelberg-New York, 1984.
19. Ciganek, E. Org. React., **1984,** *32,* 1.
20. Fallis, A.G. *Canad. J. Chem.* **1984,** *62,* 183.
21. Primeau, J.L.; Anderson, R.C.; Fraser-Reid, B. *J. Am. Chem. Soc.* **1983,** *105,* 5874.
22. Horton, D.; Machinami, T. *J. Chem. Soc., Chem. Commun.* **1981,** 88.
23. Horton, D.; Machinami, T. *Carbohydr. Res.* **1983,** *121,* 135.
24. Franck, R.W.; John, T.V. Olejniczak, K. *J. Am. Chem. Soc.* **1982,** *104,* 1106.
25. Franck, R.W.; John, T.V. *J. Org. Chem.* **1980,** *45,* 1170.
26. Giuliano, R.M.; Buzby, J.H. *Carbohydr. Res.* **1986,** *158,* C1.
27. Lopez, J.C.; Lameignere, E.; Lukacs, G. *J. Chem. Soc., Chem. Commun.* **1988,** 514.
28. Lopez, J.C.; Lameignere, E.; Lukacs, G. *J. Chem. Soc., Chem. Commun.* **1988,** 706.
29. Tejima, S.; Ness, R.K.; Kaufmann, R.L.; Fletcher, H.G. *Carbohydr. Res.* **1968,** *7,* 485.
30. Herczegh, P.; Zsély, M; Szilágyi, L.; Bognár, R. *Tetrahedron Lett.* **1988,** *29,* 481.
31. Herczegh, P.; Zsély, M.; Szilágyi, L.; Batta, Gy.; Bajza, I.; Bognár, R. *Tetrahedron* **1989,** *45,* 2793.
32. Tejima, S.; Fletcher, H.G. *J. Org. Chem.* **1963,** *28,* 2997.
33. Herczegh, P.; Zsély, M.; Szilágyi, L.; Dinya, Z.; Bognár, R. *Tetrahedron* **1989,** *45,* 5995.
34. Borch, R.F.; Evans, A.J.; Wade, J.J. *J. Am. Chem. Soc.* **1975,***97,* 6282.
35. Borch, R.F.; Evans, A.J.; Wade, J.J. *J. Am. Chem. Soc.* **1977,** *99,* 1612.

RECEIVED January 15, 1992

Chapter 9

Carbohydrates as Chiral Templates in Stereoselective [4 + 2] Cycloaddition Reactions

Horst Kunz, Bernd Müller, Waldemar Pfrengle, Karola Rück, and Wolfgang Stähle

Institut für Organische Chemie, Johannes-Gutenberg-Universität Mainz, Becher-Weg 18–20, D–6500 Mainz, Germany

Carbohydrates are inexpensive and renewable natural products which contain numerous functional groups and chiral centers. Utilizing their pronounced complexing abilities and their content of chiral information, carbohydrates are applied as chiral auxiliaries in Diels-Alder and aza-Diels-Alder reactions. The method involves the use of Lewis acid catalysts with different complexing properties. Thus, dienes of low reactivity can be transformed to their cycloadducts with high asymmetric induction. The use of N-glycosyl imines as the dienophiles offers stereoselective access to enantiomerically pure piperidine derivatives, e.g. the alkaloids coniin and anabasin.

Carbohydrates are interesting raw material for chemical synthesis. They are a renewable source and biologically degradable. Also, from the stereochemical point of view, carbohydrates possess attractive properties. Within one molecular unit they contain numerous functional groups and chiral centers. Logically, carbohydrates have been used extensively as starting materials in chiral-pool syntheses of natural products (1,2). In contrast, until the mid eighties,when we started our investigations (3), the potential of carbohydrates as removable chiral auxiliaries in stereoselective constructions of new chiral compounds had been considered only in a few isolated cases. In this sense, reducing agents, in particular complex hydrides, have been modified by addition of carbohydrate derivatives as ligands (4). In a few other reports, asymmetric aldol reactions of carbohydrate-linked enolate components have been described, which, however, resulted in only moderate stereoselections (5, 6). Also, cycloaddition reactions have been reported, in which carbohydrate-linked dienes served as the stereoselecting components in syntheses of pyranoid compounds (6) as well as in Diels-Alder reactions (8). 1,3 Dipolar cycloadditions of N-mannofuranosyl nitrones (9) and asymmetric hetero-Diels-Alder reactions of nitroso derivatives of mannofuranose have also been

0097–6156/92/0494–0131$06.00/0

carried out (10). In the meantime, some of these processes have successfully been optimized, e.g. the asymmetric reduction of ketones by application of 9-BBN in the presence of carbohydrates (11). Furthermore, carbohydrates have also been tested as the chiral tools in some other reactions, for instance, as optically active alkylating reagents (12,13), as enol ether components in Diels-Alder reactions (14), as glycosyloxy dienes in Diels-Alder reactions in organic solvents (15) or in water (16), as carbonyl components in Paterno-Büchi reactions (17) and as N-glycosyl nitrones in the synthesis of amino phosphonic acids (18) and 1,3-dipolar cycloadditions (19). In some of these processes, double stereodifferentiation has been achieved. The optical inductions in most of the reactions are only moderate.

Our interest in this field was initiated by the appearance of crucial problems in the chemical synthesis of biologically interesting glycopeptides (20). There is a particular difficulty arising in the synthesis of O-glycosyl serine and threonine derivatives which consists of the facile base-catalyzed β-elimination of the carbohydrate part from these compounds (21). At first sight this base-sensitivity of O-glycosyl serine and threonine derivatives is surprising, since tert-butyl and benzyl ethers of these amino acids and their esters are stable towards bases and, thus, are frequently utilized in peptide chemistry. When we became confronted with these contradictory results from the two divisions of natural product chemistry, carbohydrate chemistry on the one hand and peptide chemistry on the other, we were forced to explain this phenomenon. Thus, we rationalized the high leaving group quality of the carbohydrate part of O-glycosyl serine and threonine derivatives in terms of their pronounced complexing abilities towards cations.

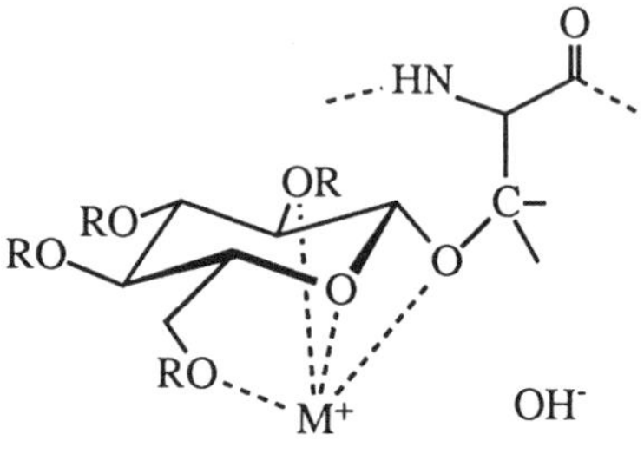

A

The complexation in **A** increases the leaving group potential of the carbohydrate by a pre-arranged neutralization of the escaping anomeric anion. In addition, the elimination might also be favored by the higher local concentration of counter-anions in the neighbourhood of the coordinatively fixed cation.

This interpretation of the base-catalyzed β-elimination of the glycan part from glycosyl serine and threonine structures, in turn, initiated the idea to utilize these complexing abilities of the polyfunctional carbohydrates together with their high chiral content for the design of stereoselective reactions (3). The application of this concept to cycloadditions is outlined in this chapter.

Lewis Acid-catalyzed Diels-Alder Reactions

Lewis acid catalysis plays a key role in stereoselective Diels-Alder reactions (22). It allows the reactions to be run at low temperature and often enforces an efficient chelation control. Carbohydrates carry a number of functional groups located at chiral centers. Logically, both the dienophile (or diene) component and the Lewis acid catalyst can be covalently bound to a single monosaccharide unit. In this sense, the 3-O-acryloyl glucofuranose derivative **1** was in situ synthesized by reacting the corresponding 5,6-di-O-trimethylsilyl ether with titanium tetrachloride at -78°C (23). Since the proton NMR signals of the starting bis-silyl ether appear in the sequence $\delta_{\beta\text{-}Z} > \delta_{\alpha} > \delta_{\beta\text{-}E}$, the acrylate moiety of **1** apparently prefers the normal anti conformation. In the reactive complex, the intramolecular complexation of the enoate carbonyl by the titanium center pre-orients the dienophile to a conformation in which the *re* side is efficiently shielded.

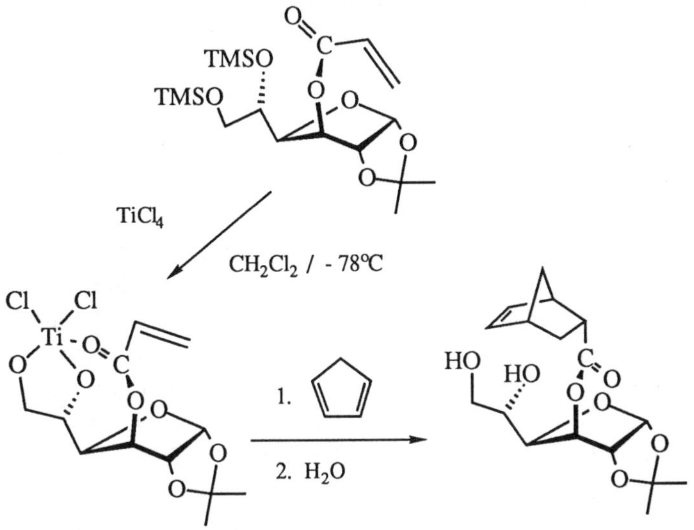

TMSO TMSO $TiCl_4$ CH_2Cl_2 / - 78°C 1. 2. H_2O

1 **2** 82 %, (R) : (S) = 93 : 7

Therefore, cyclopentadiene attacks the complex **1** with a clear preference for the unshielded *si* side. As expected, the (R)-diastereomer **2** of the cycloadduct is formed in a high diastereoselectivity (93:7). The pure major diasteromer **2** is isolated by flash-chromatography. Its reductive cleavage to give (1R, 2R) 5-

norbornen-2-yl methanol confirms the absolute configuration of the adduct **2**. The minor (1S, 2S) diastereomer presumably arises from the Diels-Alder reaction of the syn-rotamer of **1**. Since the amount of syn-rotamer increases with enhanced temperature, less reactive dienes requiring room temperature for their reaction with **1**, e. g. anthracene, give the corresponding cycloadducts with only low stereoselectivity (70 : 30).

On the contrary, if the reactivity of the dienophile complex is enhanced, a higher stereoselection can be expected. Actually, the titanium complex **3** of the 3-O-acryloyl-1,2-O-isopropylidene-α-D-xylofuranose, in which the titanium carries three chlorine ligands, with cyclopentadiene forms only one diastereomer **4** of the Diels-Alder adduct. As confirmed by reductive cleavage, only the expected (R)-diastereomer is obtained. The other diastereomer is not detectable by 400 MHz-NMR spectroscopy of the crude hydrolyzed reaction mixture (23).

CH_2Cl_2
- 78°C

3 **4**

73% , exclusively (R)

Since in both these Diels-Alder reactions the titanium catalyst is covalently bound to the carbohydrate matrix as a titanate, its Lewis acidity is distinctly reduced. Only reactive dienes, e. g. cyclopentadiene, undergo cycloaddition with these complexes **1** and **3** at low temperature and with high stereoselectivity.
These limitations can be overcome by using carbohydrate templates, which allow a coordinative fixation of the Lewis acid catalysts and, furthermore, are stable even to strong Lewis acids. The 4,6-di-O-pivaloyl dihydroglucal **5** fulfils these demands. Under catalysis by Lewis acids, the acrylate **6** of the template **5** smoothly reacts with dienes of quite different reactivity (24). Depending upon the reactivity of the diene the strength of the Lewis acid catalyst can be adjusted, so that the prevailing conditions allow the cycloaddition to be carried out optimally in both efficiency and diastereoselectivity (Table I). In the presence of the appropriate titanium catalyst (one equivalent per ester carbonyl) cycloadducts **8** obtained from **6** and cyclopentadiene, cyclohexa-1,3-diene, isoprene, butadiene, anthracene and 9,10-dimethoxy-anthracene are formed with high yields and good to excellent

stereoselectivity. As is expected from the assumed intermediate complex **7**, the (R)-diastereomers of the adducts **8** are obtained as the major products.

5 → **6**

$TiCl_2(OiPr)_2$, CH_2Cl_2 /-30°C

7 → **8**

Table I: Lewis Acid-catalyzed Asymmetric Diels-Alder Reaction of the 3-O-Acryloyl Dihydroglucal Derivative **6** with Dienes of Different Reactivity in Dichloromethane

Diene 10 equiv.	Lewis Acid 3 equiv.	Reaction Temp. (°C)	Reaction Time (h)	8	Yield[a] (%)	d.r.[b] (R) : (S)
cyclpentadiene	$TiCl_2(OiPr)_2$	-30	1.5	**a**	92	98 : 2[c]
cyclohexadiene-1,3	$TiCl_4$	0	2	**b**	88	90 : 10
isoprene	$TiCl_3(OiPr)$	0	48	**c**	94	95 : 5
butadiene-1,3	$TiCl_3(OiPr)$	0	6	**d**	90	92 : 8
anthracene[d]	$TiCl_4$	0	4	**e**	87	99 : 1
9,10-di-MeO-anthracene[e]	$TiCl_3(OiPr)$	0	24	**f**	93	99 : 1

a) Purified by flash-chromatography; b) d.r. = diastereomeric ratio; c) endo/exo >30:1 (400 MHz ^{1}H-NMR); d) 2 equivalents; e) 1.5 equivalents.

The reactive cyclopentadiene affords the corresponding cycloadduct **8a** with high yield and stereoselectivity if the Lewis acid $TiCl_2(OiPr)_2$ is applied at -30°C. The same conversion of **6** to **8a** carried out at -78°C and catalyzed by the stronger

Lewis acid $TiCl_4$, results in a reduced stereoselection (d.r. 80:20). On the other hand, dienes of low reactivity, e. g. cyclohexadiene and anthracene, require activation by the strong Lewis acid $TiCl_4$ in order to furnish the cycloadducts **8b** or **8e**, respectively, with high yield and stereoselection. Dienes of moderate reactivity, e. g. isoprene and 9,10-di-methoxy-anthracene, are converted to their cycloadducts with **6** by using $TiCl_3(OiPr)$ as the catalyst. The absolute configuration of the major diastereomers **8** is confirmed by reduction of the ester **8a** to the alcohol .

The efficient stereodifferentation in cycloadditon reactions of the dihydroglucal-linked acrylate **6** can be traced back to a clear diastereofacial differentiation at the α,β-unsaturated carbonyl system arising in the complex **7**. The bulky pivaloyloxy group at C-4 boosted by complexation provides an almost perfect shielding of the *re* side of the dienophile, whereas the lacking functionality at C-2 allows a free access of the diene component to the *si* side.

The concept of carrying out stereoselective Diels-Alder reactions with dihydroglycal-linked dienophiles, offers the possibility of selectively generating the opposite enantiomeric configuration of the cycloadducts. To achieve this aim, the 4-O-pivaloyl dihydro-L-rhamnal **9** is applied as the auxiliary (24). It is formally enantiomeric to **5**. For the acrylate **10** and its titanium complex **11** the same side-differentiating factors that have been described for the glucal analogue **6** cause an efficient shielding of the *si* side. Logically, the titanium catalyzed Diels-Alder reactions of **10** selectively furnish the (S)-diastereomeric adducts **12**.

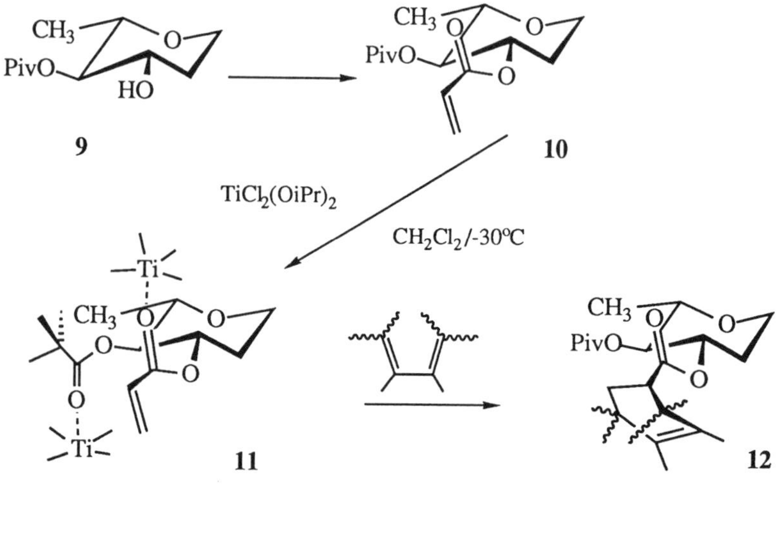

CH3 PivO HO O **9** → CH3 PivO O O O **10**

$TiCl_2(OiPr)_2$ CH_2Cl_2/-30°C

Ti CH3 O O O O O Ti **11** → CH3 PivO O O O **12**

Optimal results of these conversions again are obtained by varying the strengh of the Lewis acid depending upon the diene (Table II). The reactive cyclopentadiene requires only moderate Lewis acid catalysis [$TiCl_2(OiPr)_2$] at -30°C in order to give the (S)-diastereomeric cycloadduct **12a** in high yield and stereoselectivity. The analogous reaction of isoprene to form **12c** is favorably carried out with slightly enhanced Lewis acid catalysis [$TiCl_3$(OiPr)], whereas cyclohexadiene, butadiene and anthracene demand activation by $TiCl_4$. The corresponding adducts **12b, 12d** or **12e,** respectively, are obtained with good to excellent asymmetric induction (24). Reductive cleavage of the norbonenyl carboxylic acid ester **12a** allows the assigment of the absolute configuration of the adducts **12**. The major diastereomers revealed to be (1S, 2S)-configured.

Table II: Lewis Acid-catalyzed Asymmetric Diels-Alder Reactions of the 3-O-Acryloyl Dihydro-L-rhamnal Derivative **10** with Dienes of Different Reactivity in Dichloromethane

Diene 10 equiv	Lewis Acid 2 equiv	Reaction Temp (°C)	Reaction Time h	**12**	Yield[a] (%)	d.r.[b] (R) : (S)
cyclopentadiene	$TiCl_2(OiPr)_2$	-30	2	**a**	80	6 : 94
cyclohexadiene-1,3	$TiCl_4$	0	18	**b**	74	7 : 93
isoprene	$TiCl_3$(OiPr)	-30	18	**c**	60	10 : 90
butadiene	$TiCl_4$	-30	4	**d**	68	5 : 95
anthracene[b]	$TiCl_4$	0	18	**e**	52	1 : 99

a) After purification by flash-chromatography; b) 2 equivalents.

While the cycloaddition reactions with both acrylates **6** and **10** are efficient and proceed with high stereoselectivity, the analogous reactions of the corresponding crotonoates give only unsatisfying results. Asymmetric Diels-Alder reactions of crotonoates have been described, for instance, by Evans et al. (25) who used an oxazolidinone derived from phenylalaninol as the chiral auxiliary. With the aim of developing stereoselective Michael-type reactions consisting of the 1,4 addition of dialkylaluminium chlorides to α,β-unsaturated N-acyl amides (26), we have developed a new bicyclic oxazolidinone auxiliary derived from amino sugars (27). The N-crotonoyl derivative **13** of this carbohydrate oxazolidinone proved to be very useful in asymmetric Diels-Alder reactions. The dienophile **13** reacts with cyclopentadiene in the presence of diethylaluminium chloride as the catalyst to furnish the cycloadduct **14** with high endo/exo selectivity and excellent asymmetric induction. The major endo stereoisomer is obtained in only one diastereomeric form **14** detectable by GLC.

PivO OPiv O PivO N O CH_3 O O

1.4 eq. Et_2AlCl
-78°C, dichloromethane

PivO OPiv O PivO N O CH_3 O O

13

14

Yield: 60 %, endo : exo = 20 : 1

Endo-Component:
Ratio of Diastereomers: 99 : 1, GLC

The examples described above also illustrate the potential of carbohydrate auxiliaries in stereoselective syntheses of chiral carbocylic compounds via cycloadditions. Furthermore, their use can be extended to the stereoselective synthesis of heterocyclic compounds.

Aza Diels-Alder Reactions

Glycosylamines, in particular the O-pivaloyl protected galactopyranosylamine **15** proved to be very efficient chiral templates in asymmetric Strecker (28) and Ugi syntheses (29) of α-amino acid derivatives. In both these processes the imines **16** formed from the galactosylamine **15** and aldehydes are the real chiral substrates. The N-galactosyl imines **16** of aromatic aldehydes can be synthesized from **15** and the aldehyde in n-pentane in the presence of molecular sieves (4Å) and dried silica gel. Under these conditions anomerization of the imines can be prevented. The compounds **16** are isolated in crystalline form and contain less than 4% of the corresponding α-anomer.

The N-galactosyl imines **16** are dienophiles of relatively low reactivity. They do not react with dienes at room temperature. Even in the presence of zinc chloride in tetrahydrofuran, the conversion to cycloadducts does not occur. If the more active zinc chloride etherate in dichloromethane is applied, then isoprene undergoes aza-Diels-Alder reactions with imines **16** of aromatic aldehydes of quite different electronic properties within the aromatic ring (Table III, Ref. 30). Three diastereomers are formed. Besides the two β-anomeric N-galactosyl dehydropiperidines **17** a third component is found in small amounts. It shows a UV spectrum quite similar to those obtained from the β anomers **17**. Therefore, this by-product is considered to be an α-anomer corresponding to **17**. Because 2,3-dimethylbutadiene, which cannot form regioisomeric Diels-Alder adducts, reacts with **16** analogously to give three products which show HPLC retention times exactly corresponding to those determined for the isoprene adducts, it is concluded that the addition of isoprene to the imines **16** proceeds with complete regioselectivity (30).

R-CHO

15 **16**

R = p-Cl-C_6H_4 **16d**

$ZnCl_2 \cdot Et_2O$
CH_2Cl_2

$ZnCl_2 \cdot Et_2O$
CH_2Cl_2

17

18 96% (incl 5% of α-anomer)

d.r. 87 : 13, pure diastereomer 71%

As the reaction with the imines **16** requires a temperature of +4°C and with the less reactive compounds even room temperature, the diastereoselectivity achieved is only moderate reaching up to a kinetic ratio of 9:1 for the 4-fluoro-phenyl derivative **17c**. However, the yield of the products **17** and **18** is excellent, and in most cases the pure major diastereomer of the chiral piperidine derivatives can be isolated in satisfactory yields by flash-chromatography (30). The absolute configuration of the major diastereomers has not yet been assigned. The analogy to the Ugi reaction performed with N-galactosyl imines **16** suggests that the zinc complexes of these compounds are attacked from the sterically less hindered *si* side, i. e., from the side of the ring oxygen. In general, the Schiff bases **16** prefer a conformation in which the ring C-O bond is almost perpendicular to the plane of the σ bonds of the imine due to the delocalization of the C=N π electrons into the σ* orbital of the ring C-O bond. The preference of this conformation is confirmed by a strong NOE between the aldimine and the anomeric proton in the NMR spectra (28,29,31).

Table III: Aza-Diels-Alder Reactions of Isoprene with N-Galactosyl Imines **16** to give Chiral 3,4-Dehydropiperidines **17** (30)

Imine	R	Product 17	Kinetic Ratio[c)]	Yield (%) Mixture	α-Anomer	Pure [d)] β-Diastereomer
16a	2-Furyl	**17a**[a)]	79 : 21	98	-	-
16b	2-Thienyl	**17b**[a)]	85 : 15	95	8	-
16c	4-F-Ph	**17c**[b)]	90 : 10	90	15	52
16d	4-Cl-Ph	**17d**[b)]	85 : 15	95	12	60
16e	3-Pyridyl	**17e**[a)]	69 : 31	98	-	48

a) Reaction temp. 20°C, time 96h; b) reaction temp. 4°C, time 12h; c) β-anomers, recorded by HPLC from hydrolyzed reaction mixture; d) isolated by flash-chromatography; e) 2 equiv. of $ZnCl_2/Et_2O$ are required for running the reaction.

It is concluded from the results of the cycloaddition reactions of isoprene with the N-galactosyl imines **16**, that the application of more electron-rich dienes should allow the reaction to be conducted at lower temperature and with higher stereo-selectivity. In this sense, the N-galactosyl-imines **16** are reacted with 1-methoxy-3-trimethylsilyloxybutadiene **19** (32) to give the N-galactosyl 2,3-dehydropiperidin-4-one derivatives **20** formed by subsequent elimination of methanol from the assumed cycloaddition products (33).

As a consequence of the higher reactivity of the silyl dienol ether **19**, the transformations already proceed at -20°C and with reduced Lewis acid catalysis by performing the reactions in the presence of one equivalent of $ZnCl_2 \cdot OEt_2$ in tetrahydrofuran.The mild reaction conditions allow to use imines of aliphatic aldehydes prone to easy anomerization, e. g. the imine **16f** of butyraldehyde, as the dienophile. After quenching with dilute HCl, the N-galactosyl dehydropiperidinones **20** are obtained in high yield and with excellent stereoselectivity. The ratios of diastereomers are determined by HPLC directly carried out with the crude product. Pure diastereomers are isolated in high yield by flash-chromatography (**20a**/**20d**) or recrystallization (**20f**).

PivO OPiv O PivO PivO N=C(H)-R + H_3CO–CH=CH–C(=CH$_2$)OSiMe$_3$ **19** —($ZnCl_2\,Et_2O$, THF)→ PivO OPiv O PivO PivO N (ring) =O H R

				(R) : (S)	Yield	Pure **20**
16a	R =	2-furyl	**20a**	96 : 4	95%	60%
16d	R =	Cl–C$_6$H$_4$–	**20d**	98 : 2	95%	90%
16f	R =	n-C_3H_7	**20f**	2.5 : 97.5	96%	81%

The assigment of the absolute configuration of the products **20** is achieved by conversion of the n-propyl derivative **20f** to the alkaloid coniin. To this end, the C=C double bond of **20f** is reduced with L-selectride (34). The resulting piperidinone is converted to its dithiolane derivative which subsequently is subjected to desulfurization with Raney nickel to deliver the N-galactosyl coniin **21**. The final release of the enantiomerically pure alkaloid **22** from the carbohydrate template is achieved by treatment of **21** with HCl/methanol (31).

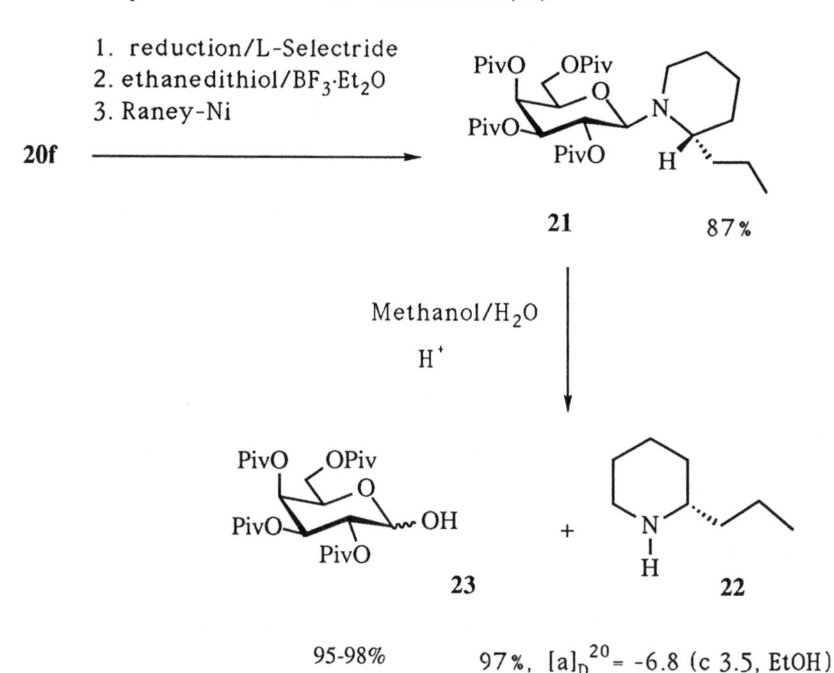

During work up of the hydrolyzed reaction mixture, the carbohydrate matrix **23** can be re-isolated almost quantitatively. Coniin **22** is obtained in enantiomerically pure form in the natural (S)-configuration. This experiment confirms the absolute configurations given for the compounds **20**. However, these results do not indicate the absolute configuration of the cycloadducts **17** obtained from isoprene, since the pyridine aldimine **16e** reacts with the silyloxy butadiene **19** to form the corresponding N-galactosyl dehydropiperidinone **24** with opposite configuration compared with compounds **20f** (30, 33). The reaction requires the application of 2 equivalents of $ZnCl_2 \cdot Et_2O$, but then proceeds very effective and delivers the (S)-configurated product **24** with a diastereomeric ratio of 96:4.

$OSiMe_3$ H_3CO **19**

16e

2 equiv. $ZnCl_2 \cdot Et_2O$
THF

PivO OPiv O PivO PivO N O H N

24

> 90% (S) : (R) = 24 : 1

1. Reduction / L-Selectride
2. Ethanedithiol/$BF_3 \cdot Et_2O$
3. Raney-Ni

PivO OPiv O PivO OH PivO + HN H N

Methanol/H_2O
H^+

PivO OPiv O PivO PivO N H N

23 **26** **25**

The assigment of the absolute configuration of **24** is achieved by its transformation to the N-galactosyl anabasin **25** and the final releasing of the enantiomerically pure (S)-anabasin **26** from the carbohydrate template. The obtained anabasin shows an optical rotation value $[[\alpha]^{20}_D = -130.8$ (c = 1.2, MeOH)] identical with that reported in the literature (35) for the enantiomerically pure (S)-anabasin.

To clarify the opposite course of the reaction of the silyl dienol ether **19** with the N-galactosyl imines **16a,d,f** on the one hand and the 3-pyridyl derivative **16e** on the other, the formations of **20** and **24** have been investigated in greater detail. These closer examinations revealed that in the presence of zinc chloride the silyl dienol ether **19** does not react in a concerted cycloaddition pathway with the N-galactosylimines **16**. Instead, the conversion to the apparent cycloadducts turned out to be a tandem Mannich-Michael reaction sequence followed by a condensation step. If these reactions are quenched with aqueous ammonium chloride solution, the primarily formed Mannich bases **27** from **16a** and **16d** or **28** (from pyridine aldimine **16e**, respectively) can be isolated and characterized (33). The diastereomeric ratios measured for the Mannich bases **27** or **28**, respectively, are identical to those found for the cyclic products **20** and **24**. On treatment with 1N HCl, the Mannich bases **27** and **28** rapidly undergo hetero Michael addition and, subsequently, irreversible condensation reaction to furnish the corresponding N-galactosyl dehydropiperidinones **20** or **24**, respectively, which again show ratios of diastereomers identical to those recorded for **27** and **28**.

16a,d → (H_3CO–CH=CH–C($OSiMe_3$)=CH_2 **19**; 1. 1 equiv. $ZnCl_2 \cdot Et_2O$, THF / -20°C; 2. aq. NH_4Cl)

27a R = 2-Furyl, 36% (+ 60% **17a**)

27d R = 4-Cl-Phenyl, 82%

16e → (H_3CO–CH=CH–C($OSiMe_3$)=CH_2 **19**; 1. 2 equiv. $ZnCl_2 \cdot Et_2O$, THF / - 20°C; 2. aq. NH_4Cl)

28 57%

The very effective stereoselection, which is the diastereofacial differentiation occuring during the initial Mannich reaction, can be rationalized by the complexing and chiral properties of the carbohydrate templates. The opposite, but in both cases excellent, stereoselection found for the imines **16a-d,f** on the one hand and the 3-pyridyl derivative **16e** on the other, is due to the different push-pull activation of the reaction components as is illustrated in the Formulas **B** and **C**. The push activation of the silyl dienol ether by zinc chloride complexes of usual aldimines **16** can only occur in front of the imine double bond. Therefore, the C-nucleophile **19** in these cases is introduced from the front side (Formula **B**) and delivers selectively Mannich bases of configuration **27**.

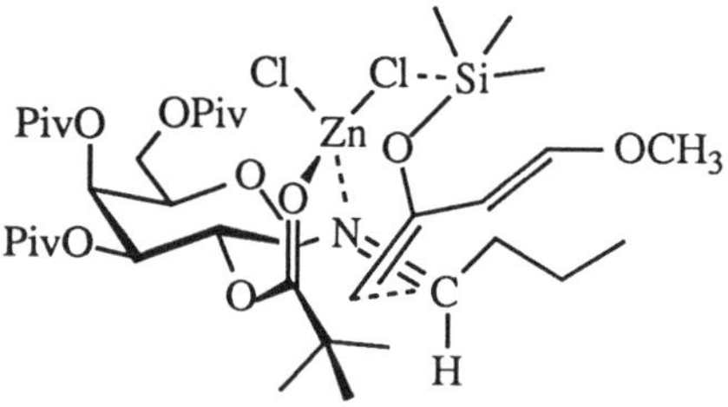

B 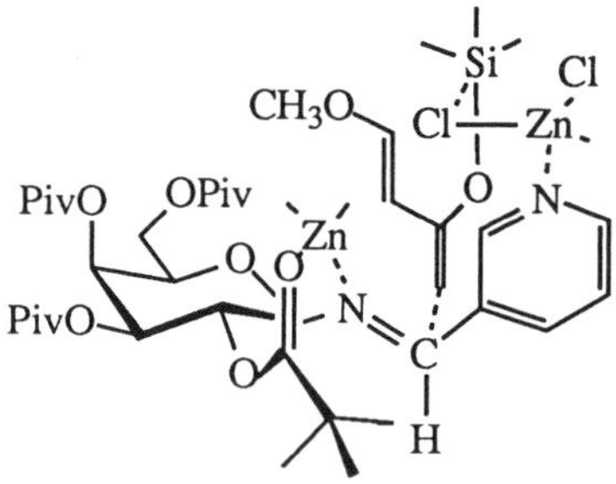 **C**

For the reaction of the 3-pyridine aldimine **16e**, the first equivalent of zinc chloride is coordinated to the more nucleophilic pyridine nitrogen and thus is inactivated. Only the second equivalent activates the imine. However, the zinc chloride coordinated to the more nucleophilic pyridine nitrogen, logically exposes more nucleophilic chlorine ligands for the interaction with the silyl dienol ether **19.** Therefore, in this case the nucleophile **19** is introduced from the free back side of **16e** (Formula **C**) and furnishes selectively the Mannich base **28** with opposite stereochemistry compared to the compounds **27**.

In conclusion, these studies further demonstrate that the carbohydrates are plentiful sources of inexpensive chiral auxiliaries. On one molecular unit they carry a high density of functional and chiral information which can be amplified by complexation with Lewis acidic centers. Taking advantage of these properties the carbohydrates can efficiently be used as the chiral templates in [4+2] cycloaddition reactions and also in other types of chemical transformations.

Acknowledgement. This work was supported by the "Deutsche Forschungsgemeinschaft", the "Bundesminister für Forschung und Technologie" and the "Fonds der Chemischen Industrie".

Literature Cited

1. For a review, see: Hanessian, S. *Total Synthesis of Natural Products: The Chiron Approach;* Pergamon Press: Oxford, U.K. 1983.

2. For recent applications, see: Mulzer, J.; Büttelmann, B.; Münch, W. *Liebigs Ann. Chem.* **1988**, 445. and references cited therein.
3. Kunz, H. in *Selectivities in Lewis acid-promoted reactions,* Schinzer, D., Ed.; Kluwer Publ.: Amsterdam, The Netherlands **1989**, 189 - 202.
4. Landor, S. R.; Miller, B. J.; Tatchell, A. R. *J. Chem. Soc.* C **1966**, 2280.
5. Brandänge, S.; Josephson, S.; Mörch, L.; Vallén, S. *Acta Chem. Scand.* B **1981**, *35*, 1296.
6. Heathcock, C. H.; White, C. T.; Morrison, J. J.; Van Derveer, D. *J. Org. Chem.* **1981**, *46,* 1296.
7. David, S.; Eustache, J.; Lubineáu, A. *J. Chem. Soc.* Perkin Trans I **1973**, 2274.
8. Gupta, R. C.; Slavin, A. M.; Stoodley, R. J.; Williams, D. J. *J. Chem. Soc. Chem. Commun.* **1986**, 668.
9. Vasella, A.; Voeffray, R. *Helv. Chim. Acta* **1982**, *65*, 1953.
10. Felber, H.; Kresze, G.; Prewo, R.; Vasella, A. *Helv. Chem. Acta* **1986**, *69*, 1137.
11. Brown, H. C.; Cho, B. T.; Park, W. S. *J. Org. Chem.* **1986**, *51*, 1934 and references cited therein.
12. Duhamel, P.; Eddine, J. J.; Valnot, J.-Y. *Tetrahedron Lett.* **1987**, *28*, 3801.
13. Umimura, K.; Matsuyama, H.; Kobayashi, M.; Kamigata, N. *Bull. Chem. Soc. Jpn.* **1989**, *62*, 3026.
14. Choudhury, A.; Franck, R. W.; Gupta, R. B. *Tetrahedron Lett.* **1989**, *30*, 4921.
15. Larsen, D. S.; Stoodley, R. J. *J. Chem. Soc.* Perkin Trans I **1990**, 1339.
16. Lubineáu A.; Queneáu, Y. *Tetrahedron* **1989**, *45*, 6697.
17. Pelzer, R.; Jütten, P.; Scharf, H.-D. *Chem. Ber.* **1989**, *122*, 487.
18. Huber, R.; Vasella, A. *Helv. Chim. Acta.* **1987**, *70*, 1461.
19. Mzengeza, S.; Whitney, R. A. *J. Org. Chem.* **1988**, *53*, 4074.
20 Kunz, H. *Angew. Chem. Int. Ed. Engl.* **1987**, *26*, 294.
21. For a review, see: Marshall, R. D.; Neuberger, A. *Adv. Carbohydr. Chem. Biochem.* **1970**, *25*, 407.
22. For a review see: Helmchen, G.; Karge, R.; Weehman, J. in *Modern Synthetic Methods*, Sheffold, R., Ed., Springer-Verlag: Heidelberg, Germany, **1986**, Vol. 4, p. 262.
23. Kunz, H.; Müller, B.; Schanzenbach, D. *Angew. Chem. Int. Ed. Engl.* **1987**, *26*, 267.
24. Stähle, W.; Kunz, H. *Synlett* **1991**, 260.
25. Evans, D. A.; Chapman, K. T.; Bisaha, J. *J Am. Chem. Soc.* **1984**, *106*, 4261.
26. Rück, K.; Kunz, H. *Angew. Chem. Int. Ed. Engl.* **1991**, *30*, 694, and references cited therein.

27. Kunz, H.; Rück, K. *German Patent Appl.* P 4113327.7, 24. April 1991.
28. Kunz, H.; Sager, W. *Angew. Chem. Int. Ed. Engl.* **1987**, *26*, 557.
29. Kunz, H.; Pfrengle, W. *J. Am Chem. Soc.* **1988,** *110*, 651.
30. Pfrengle, W.; Kunz, H. *J. Org. Chem.* **1989,** *54*, 4291.
31. Kunz, H.; Sager, W.; Schanzenbach, D.; Decker, M. *Liebigs Ann. Chem.* **1991**, 649.
32. Danishefsky, S.; Langer, M. E.; Vogel, C. *Tetrahedron Lett.* **1985**, *26,* 5983, and references cited therein.
33. Kunz, H.; Pfrengle, W. *Angew. Chem. Int. Ed. Engl.* **1989**, *28*, 1067.
34. Brown, H. C.; Krishnamurthy, *J. Am. Chem. Soc.* **1972**, *94*, 7159.
35. Späth, E.; Kesztler, F. *Ber. Dtsch. Chem. Ges.* **1937**, *70*, 704.

RECEIVED December 2, 1991

Chapter 10

Aqueous Cycloadditions Using Glycoorganic Substrates

Stereo- and Physicochemical Aspects

André Lubineau, Jacques Augé, Hugues Bienaymé, Nadège Lubin, and Yves Queneau

Laboratoire de Chimie Organique Multifonctionelle, Institut de Chimie Moléculaire d'Orsay, Université Paris-Sud, Bât. 420, F–91405 Orsay Cedex, France

Water as solvent is shown to enhance both the rate and the selectivity in Diels-Alder reactions using several water-soluble *trans* buta-1,3-dienyl derivatives of carbohydrates from the simplest one with only one asymmetric center, derived from D-glyceraldehyde to partially protected buta-1,3-dienyl glucosides. In each case, virtually complete *endo* selectivity was observed, and it is shown that the facial stereoselectivity can be modulated through chemical manipulation of the sugar moiety. It is also shown that the rate enhancement in such aqueous cycloadditions has an entropic origin.

In the early seventies, we demonstrated (*1*) that carbohydrate-derived buta-1,3-dienyl ethers are good precursors for the preparation of oligosaccharides. This methodology reached its highest point with the synthesis of the epitope of the blood group antigen A using the buta-1,3-dienyl ether **1** which gave a 97:3 facial stereoselectivity toward n-butyl or (-) menthylglyoxylate (*2*). So, when in 1980, Breslow *et al.* (*3*) showed that water as solvent enhances both the rate and the *endo* selectivity in the Diels-Alder reaction between cyclopentadiene and but-2-enone, we decided to investigate our methodology in these newly described conditions. However, it was obvious that to take advantage of the hydrophobic effect, the reactants need to be, at least partially, water-soluble. For preparative purposes in the field of quassinoids, Grieco *et al.* (*4*) devised a water-soluble diene synthesis using a carboxylate group as the hydrophilic part. However, from a general point of view, if not needed, this functionnal group suffers from the fact that it forms a permanent substituent in the product. We decided to synthesize buta-1,3-dienyl ethers of unprotected sugars (*5,6*) without knowing at that time whether the hydrophilic part would not prevent the hydrophobic one from the expected entropy-driven association. In contrast to what we had done before, we chose first to link the buta-1,3-dienyl moiety to the anomeric position, anticipating that the sugar could give water solubility and act as a chiral template. An additionnal consideration was the possibility that the sugar moiety could be removed after the aqueous cycloaddition by acidic or even at neutral pH by enzymatic means. We describe in this paper the preparation of various carbohydrate-derived buta-1,3-dienes from the

0097–6156/92/0494–0147$06.00/0

simplest unactivated one, with only one asymmetric carbon atom derived from D-glyceraldehyde to variously protected buta-1,3-dienyl glucosides and the stereochemical course of their cycloadditions with standard dienophiles. We also include some generalizations about the thermodynamics of such aqueous cycloadditions.

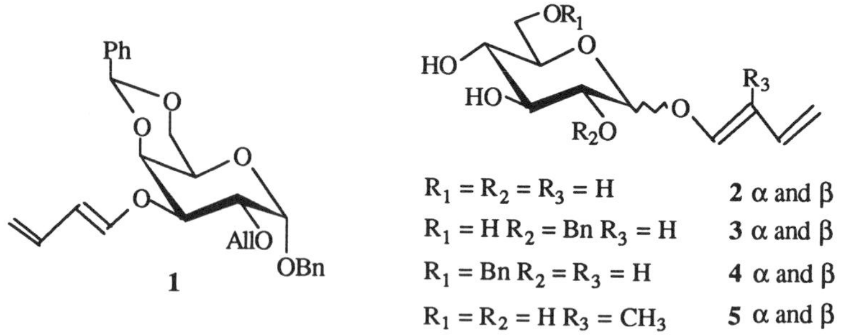

1

$R_1 = R_2 = R_3 = H$ **2** α and β
$R_1 = H$ $R_2 = Bn$ $R_3 = H$ **3** α and β
$R_1 = Bn$ $R_2 = R_3 = H$ **4** α and β
$R_1 = R_2 = H$ $R_3 = CH_3$ **5** α and β

Preparation of Buta-1,3-dienyl Glucosides.

We have prepared *(6,7)* eight different buta-1,3-dienyl glucosides **2-5** α and β to study the origin of the observed facial stereoselectivity in the cycloadditions. The parent dienes 2α and 2β were synthesized as depicted in Scheme 1.

Scheme 1. Syntheses of dienes **2** and **5**, α and β

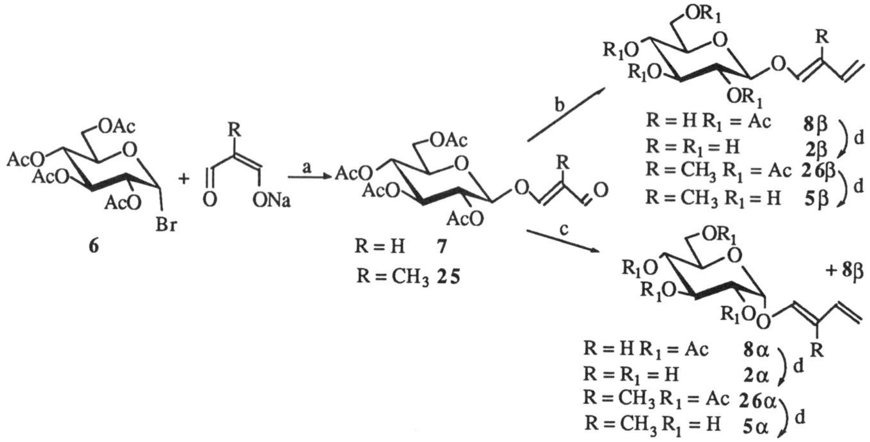

6

$R = H$ **7**
$R = CH_3$ **25**

$R = H$ $R_1 = Ac$ **8β**
$R = R_1 = H$ **2β**
$R = CH_3$ $R_1 = Ac$ **26β**
$R = CH_3$ $R_1 = H$ **5β**

$R = H$ $R_1 = Ac$ **8α**
$R = R_1 = H$ **2α**
$R = CH_3$ $R_1 = Ac$ **26α**
$R = CH_3$ $R_1 = H$ **5α**

(a) DMSO, R =H : 56%, R = CH_3 : 48% ; (b) $Ph_3P{=}CH_2$, THF-Toluene, -78°C, R = H : 83%, R = CH_3 : 74% ; (c) $Ph_3P{=}CH_2$, THF-Toluene, 20°C, R =H :72%, **8α**:**8β**, 3/1, R = CH_3 : 82%, 26α:26β, 1/2.5; (d) $MeOH\text{-}NEt_3\text{-}H_2O$, 8:1:1, 100%.

The key step is the reaction of the sodium salt of malonaldehyde with acetobromoglucose which gave the β-unsaturated aldehyde **7** in 56% yield by direct crystallization without chromatography. Then, Wittig reaction with methylene triphenylphosphorane conducted at low temperature (-78°C) gave only the diene **8β** whereas the reaction conducted at room temperature gave a separable mixture of **8α** and

8β in a 3:1 ratio in favor of the α anomer. In fact, this anomerization during the Wittig reaction allowed us to have access to both anomers which give reverse facial stereoselectivities (*vide infra*). The preparations of the corresponding 2-O-benzyl dienes **3α** and **3β** are described in Scheme 2. The buta-1,3-dienyl part was introduced as for dienes **2** onto the activated 2-O-benzyl glucose **14** prepared from the known 1,3-di-O-allyl-α,β-D-glucopyranose (*9*).

Scheme 2. Syntheses of 2-O-benzyl dienes

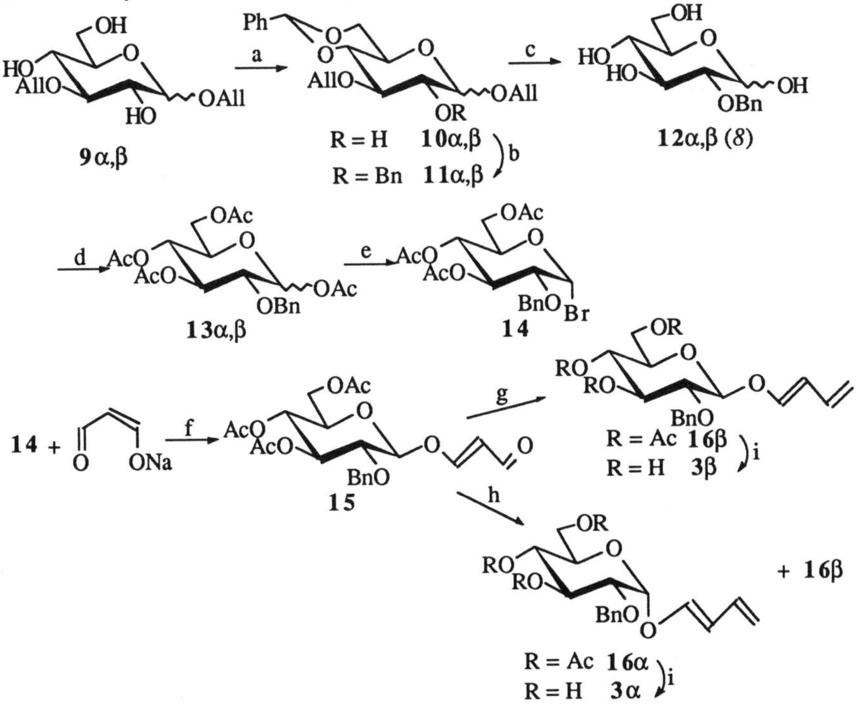

(a) PhCHO, $ZnCl_2$, 69% ; (b) BnBr, NaH, THF, 91% ; (c) tBuOK, DMSO then H_2SO_4 (2.5N) aq. acetone, 65% ; (d) Ac_2O, Pyr., 92% ; (e) HBr, AcOH ; (f) DMSO, 51% from **13α,β** ; (g) $Ph_3P{=}CH_2$, THF-Toluene, -78°C, 91% ; (h) $Ph_3P{=}CH_2$, THF-Toluene, 20°C, 93%, **16α:16β**, 1.5/1 ; (i) MeOH-NEt_3-H_2O, 8:1:1, 100%.

The 6-O-benzyl protected dienes **4α** and **4β** were prepared from the known (*10*) 1,2-O-isopropylidene-α-D-glucofuranose (Scheme 3). In this case, evidence for the intermediate formation of aldehyde **23α** in the Wittig reaction conducted at room temperature was obtained by comparison with an authentic sample of aldehyde **23α** obtained directly from compound **21** (α + β) using β-tosyloxy acrolein prepared *in situ* from the sodium salt of malonaldehyde and tosyl chloride in THF. Finally, the methyl substituted derivatives **5α** and **5β** were prepared as for the parent dienes **2α** and **2β** according to Scheme 1 using the corresponding 2-methyl substituted sodium salt of malonaldehyde (*12*).

Scheme 3. Syntheses of 6-O-benzyl dienes

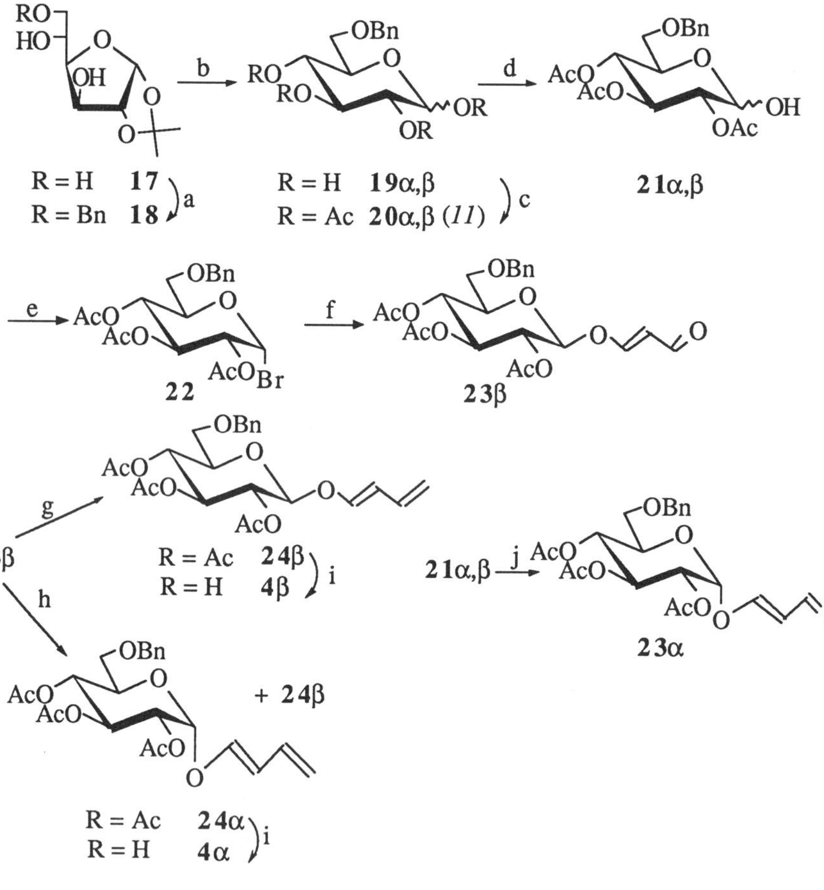

RO HO O OH O O b RO RO OBn O OR OR d AcO AcO OBn O OH OAc

R = H 17 a R = Bn 18 R = H 19α,β R = Ac 20α,β (*11*) c 21α,β

e AcO AcO OBn O AcO Br 22 f AcO AcO OBn O O O AcO 23β

AcO AcO OBn O O AcO g 23β R = Ac 24β R = H 4β i 21α,β j AcO AcO OBn O AcO O O 23α

h AcO AcO OBn O AcO O + 24β

R = Ac 24α R = H 4α i

(a) Bu_2SnO, BnBr, $Br^-N^+Bu_4$, Toluene, 77% ; (b) DOW 50 (H^+), 83% ; (c) Ac_2O, Pyr., 92% ; (d) NH_2NH_2, AcOH, DMF, 60°C, 82% ; (e) $(COBr)_2$, DMF, CH_2Cl_2 ; (f)DMSO, $NaCH(CHO)_2$, 34% from **21α,β** ; (g) $Ph_3P{=}CH_2$, THF-Toluene, -78°C, 70% ; (h) $Ph_3P{=}CH_2$, THF-Toluene, 20°C, 80%, **24α:24β**, 1.8/1 ; (i) MeOH-NEt_3-H_2O, 8:1:1, 100% (j) TsO-CH=CH-CHO, HNa, THF, 79%, **23α:23β**, 2/1.

Stereochemical Course of the Aqueous cycloaddition of Dienes 2-5, α and β with Standard Dienophiles.

Cycloadditions of the Parent Dienes 2α and 2β. The results are summarized in Table I. First, it must be noted the considerable rate enhancements of the aqueous cycloadditions using the unprotected dienes **2** (3.5 hrs at room temperature with methacrolein) were observed when compared with the corresponding cycloadditions conducted in toluene with the peracetylated dienes **8** (168 hrs at 80°C with the same dienophile). Second, a virtually complete *endo* stereoselectivity was obtained in water with **2β**, whereas this *endo* selectivity was eroded for the diene **2α**, and especially in toluene with dienes **8α** or **8β**. Third, the facial stereoselectivities correspond to addition of the dienophile to the*re* face of the diene **2β** or **8β**, and to the *si* face with dienes **2α** and 8α, regardless of the dienophiles and the solvents.

Table I. Cycloadditions of the dienes **2α** and **2β** with $(R_1)(R_2)CH=CH_2$

dienes	R_1	R_2	Temp.°C	time, hr	solvent	yield,%	*re:si*	endo:exo
2β	CH_3	CHO	20	3.5	H_2O	90	60:40	100:0
2β	H	CHO	20	3	H_2O	92	58:42	100:0
2β	H	CO_2Me	60	18	H_2O	66	73:27	100:0
2α	CH_3	CHO	20	6	H_2O	78	36:64	93:7
8β	CH_3	CHO	80	168	Toluene	80	62:38	87:13
8α	CH_3	CHO	80	168	Toluene	94	35:65	84:16

SOURCE : Adapted from Ref. 6

The facial selectivity observed with the β dienes could be explained by an approach of the dienophile from the top side of the more stable extended conformation **2β**(ext) in which a 1,3-diaxial interaction with the O-C5 bond is avoided (Figure 1). This selectivity was also observed by Stoodley *et al.* in a similar case (*13*). For the same reasons, in the case of the α-dienes **2α** and **8α**, the major approach comes from the bottom side (from the side of OH-2) which now corresponds to the *si* face of the extended form **2α**(ext) as depicted in Figure 2. These assumptions were supported using the partially protected dienes **3-5** α and β as shown below.

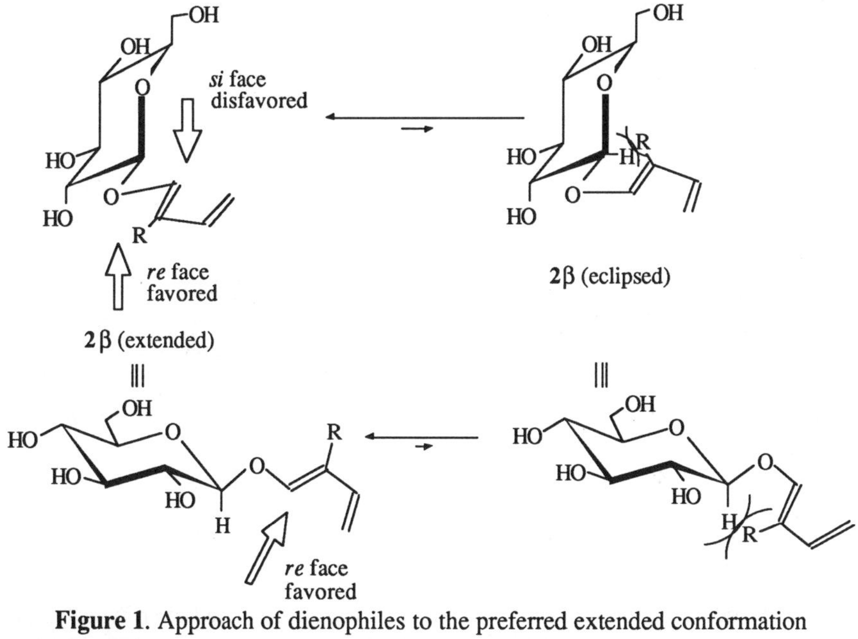

Figure 1. Approach of dienophiles to the preferred extended conformation of β-dienes.

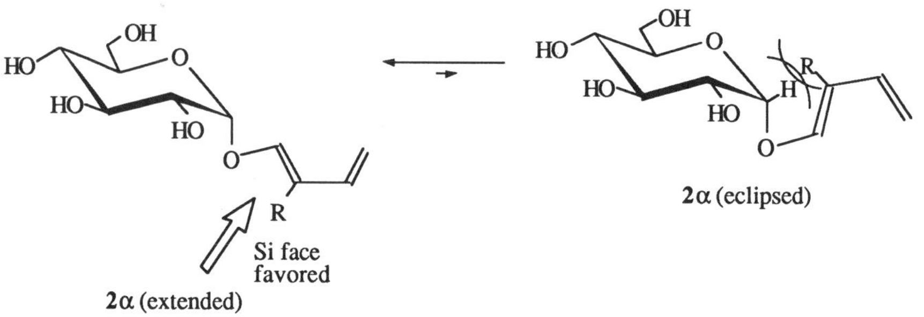

Figure 2. Approach of dienophiles on the preferred extended conformation of α-dienes.

Cycloadditions of Dienes 3-5 α and β. The 2-O-benzyl protected diene **3** with a bulky group on the side of the previously favored approach gave the reverse facial stereoselectivity *i.e.* the *si* face on the β anomer and the *re* face on the α anomer, whereas a 6-O benzyl group on the side of the already disfavored approach enhances the facial selectivity (Table II). Moreover, the contribution of the two conformers with extended or eclipsed conformations could be measured using the 2'-methyl derivatives **5α** and **5β** in which the eclipsed forms must be strongly disfavored (Figure 1 and 2). In fact, the best result was obtained with the diene **5α** which gave a 82:18 ratio of diastereoisomers in 92% yield. It must be noted that in all our cycloadditions, the major stereoisomer could always be obtained in pure form, most of the time by direct crystallisation of the peracetylated adducts. In other hand, we have also shown that we were able to cleave the glycosidic bond after the aqueous cycloadditions at neutral pH using commercially available glucosidase. This allowed us to prepare new compounds in pure enantiomeric form such as compound **27** ($[\alpha]_D^{20}$ + 15°, c O.6 CH_2Cl_2).

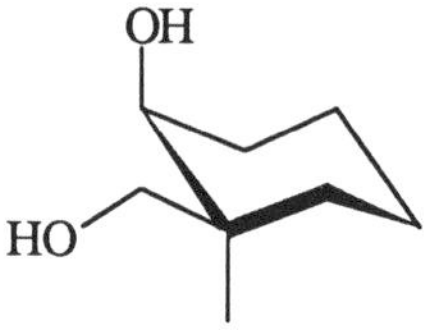

27

TableII. Cycloadditions of dienes **3-5** α and β with methacrolein

Dienes	Temp.,°C	Time, hr	Yields, %	*re:si*	*endo:exo*
3β	40	17	95	34:66	93:7
4β	40	17	86	69:31	95:5
5β	20	20	96	69:31	96:4
3α	40	17	97	57:43	93:7
4α	40	17	85	28:72	95:5
5α	20	20	92	18:82	95:5

SOURCE : Adapted from Ref. 7

The absolute configuration of the adducts were determined after hydrolysis of the glucosidic bond, reduction of the cyclohexene moiety and comparison when possible with already known compounds (as in the case of the cycloadditions with acrolein or methyl acrylate). In the case of methacrolein, the absolute configuration of **27** was determined using the Brewster procedure (*14*) which gives good approximation of the molecular rotation in cyclic compounds.

Preparation of (2S), (3E)-hexa-3,5-diene-1,2-diol 29.

This diene can be considered as the simplest possible one derived from carbohydrates as it contains only one asymmetric carbon atom provided by D-glyceraldehyde (Scheme 4). Wittig reaction between the semi-stabilized allylidenetriphenylphosphorane prepared from the corresponding phosphonium salt by deprotonation with BuLi in a THF-hexane mixture, and the 2,3-isopropylidene-D-glyceraldehyde (*15*) gave the protected diene **28** as a mixture of 30:70 ratio in favor of the *Z* isomer. When this mixture was isomerized with catalytic (0.01 mol. equi.) iodine under light, a 96:4 mixture was obtained now in favor of the *E* isomer. Hydrolysis of the acetal group with Dowex 50 (H^+) resin afforded the water soluble diene **29** (*E:Z*, 96:4) (*16*).

Scheme 4. Synthesis of diene **29**

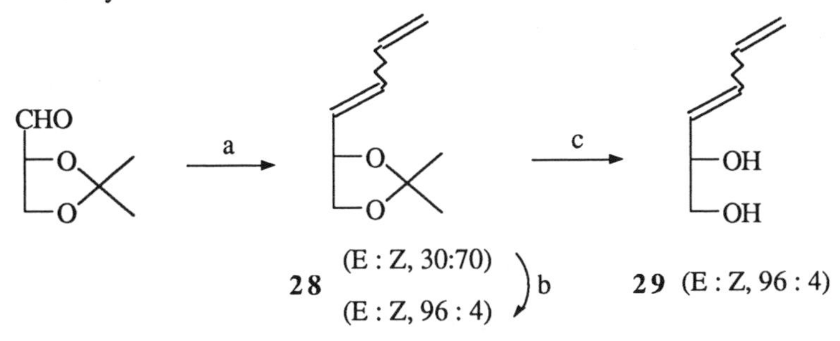

(a) $Ph_3P{=}CH{-}CH{=}CH_2$, THF-hexane, 72% ; (b) I_2, cat., hν, 94% ; (c) DOW. 50 (H^+), 100%.

Stereochemical Course of Cycloaddition Between Diene 29 and Acrolein.

As shown in Table III, there is, here again, a considerable rate enhancement in water when compared with toluene or without solvent. The aqueous cycloaddition could even be performed at 0°C which is, from the best of our knowledge, the lowest temperature used for a thermal cycloaddition with an unactivated diene. We also observed, as usual, a complete *endo* selectivity, whereas the cycloaddition of the protected diene **28** with acrolein in toluene gave rise to 17% of *exo* adducts (*16*).

Table III. Facial selectivities in the cycloadditions between the diene **29** and acrolein in various conditions

Solvent	Temp, °C	Time, hr	Yield, %	*endo si : endo re*
Toluene	60	96	94	55:45
-	60	48	73	55:45
Water	60	2	77	65:35
Water	20	24	82	65:35
Water	0	168	92	66:34

SOURCE : Adapted from Ref. 16

The facial selectivity of cycloadditions of diene **28** and **29** enabled us to assess the effect of a stereogenic allylic substituent on the stereochemical outcome of the Diels-Alder reaction (*17-19*). In fact, we obtained a mixture of the separable adducts (each as a mixture of α and β anomers) **30α,β** and **31α,β**, independantly and unambiguously analyzed by NMR spectroscopy after peracetylation as shown in Scheme 5.

In water, an increase of the facial selectivity up to 2:1 in favor of the *si* face (*anti*-addition) of the diene was observed as depicted in Figure 3. That is the contrary of what has been predicted from several models of cycloadditions. The perpendicular model in which the heteroatom is perpendicular to the dienyl system predict *syn* selectivity by *syn* periplanar approach of the dienophile (*17,18*). The coplanar model in which the heteroatom lies in the plane outside the dienyl system, predicts *syn* selectivity as well (*19*). Both of then failed to explain our *anti* selectivity without implication of the β-hydroxy group. Furthermore, the increase of the *anti*-selectivity in water compared with toluene could be a consequence of the hydrophobic effect which favored the dienophile approach *anti* to the hydrophilic face where hydroxy group are tightly bonded to water molecules.

At this point, additional question deserves some comment : why does the cycloadduct **30** adopt a pyranoid form whereas **31** prefers the furanoid one ? If we consider the oxygenated ring, compounds **30** and **31** correspond respectively to the configuration *lyxo* and *ribo* and it must be emphasized that lyxose exists exclusively in the pyranose form in aqueous solution whereas ribose partially adopts (24 %) the furanose form under the same conditions.

Scheme 5. Stereochemical outcome of the cycloaddition between the diene **29** and acrolein

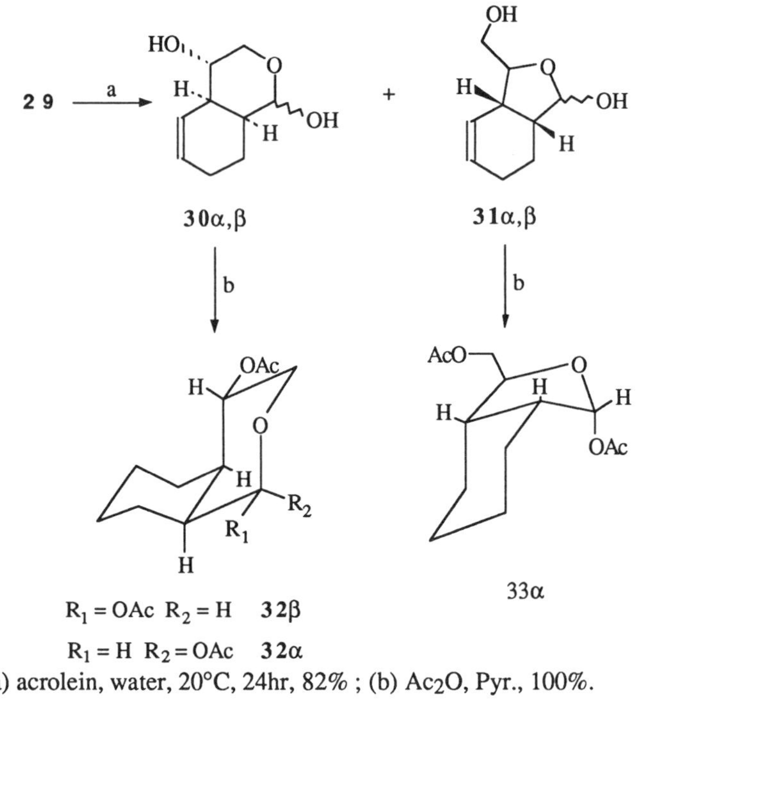

29 —a→ 30α,β + 31α,β

30α,β —b→ 32; 31α,β —b→ 33α

R_1 = OAc R_2 = H **32β**

R_1 = H R_2 = OAc **32α**

(a) acrolein, water, 20°C, 24hr, 82% ; (b) Ac_2O, Pyr., 100%.

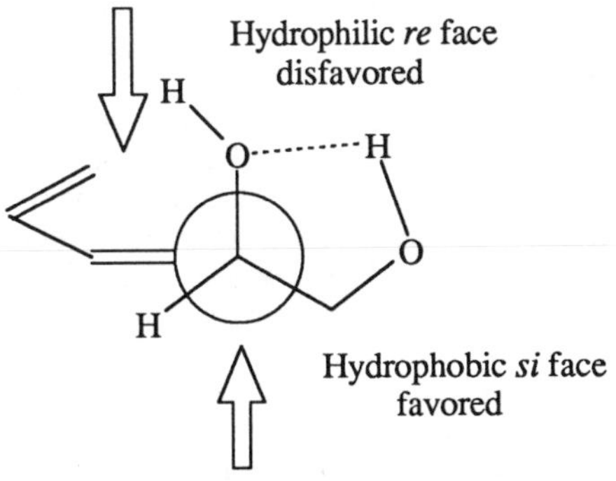

Figure 3. Approach of dienophile onto the more favored hydrophobic face.

Thermodynamics of Aqueous Cycloadditions.

In addition to the preparative aspects of aqueous cycloadditions using our glyco-organic compounds, these water soluble dienes provide insight into the thermodynamics of such reactions. In fact, since the earliest studies (*3,5,6*), the rate enhancement of aqueous cycloadditions has been associated with the hydrophobic effect, the well-known entropy-driven association of non-polar molecules or non-polar side chain groups in water which minimizes their exposure to the solvent. This concept, far from being understood on a microscopic level, has generally been associated with the three dimensional network of water molecules in the liquid state, and with the high energy necessary to create a cavity in it (*20*). To support this hypothesis, Breslow *et al.* (*3*) has shown that lithium chloride, a salting-out agent which stabilizes water structure, increases the rate of the reaction between cyclopentadiene and butadiene, whereas guanidinium chloride, a salting-in agent which decreases the hydrophobic effect by breaking-up ordered water structure, gave the opposite result. More recently, Schneider *et al.* described a quantitative correlation between solvophobicity (as measured by the standard free energy of noble gas or alkanes from gas to a given solvent) and the rate enhancement of the Diels-Alder reaction between cyclopentadiene and diethyl fumarate (*21*).

The course of other types of reactions which are also accelerated in aqueous solutions can help for a better understanding. For instance, in the Mukaiyama reaction between a silylenol ether and a carbonyl compound we obtained, in water, without any catalyst and in a similar yield, the same stereoselectivity than under 10000 atmospheres in dichloromethane that is the reverse of the normal stereoselectivity obtained by $TiCl_4$ catalysis(*22,23*). We and others have also described such a rate enhancement in the Claisen rearrangement (*24,25*). All these reactions, including the Diels Alder reaction have in common their negative activation volume. We anticipate that a reaction under kinetic control, between two small hydrophobic molecules (or between the hydrophobic moieties of amphiphilic molecules) for which the activation volume $\Delta V^{\dagger}$ is negative must be accelerated in water as it is under pressure. When several transition states are possible, the more compact should be favored. Actually, this hypothesis is verified for all our aqueous reactions, especially the Diels-Alder reactions which give virtually pure *endo* transition state which is also the transition state favored under external pressure.

We examined more closely the thermodynamics of the reaction between the diene **2β** and but-2-enone by measuring the activation parameters of the reactions conducted in water and in a water-methanol mixture (50:50, v/v, the rate of the reaction in pure methanol is too small to be recorded with reliability). To check if there is a special effect due to the presence of the sugar moiety linked to the buta-1,3-diene unit, we studied the cycloaddition between methoxybutadiene and but-2-enone under the same conditions. The results are summarized in Table IV. As anticipated, the second-order rate constant increases in pure water and the rate enhancement comes only from an increase of the activation entropy (+32.3 $J.mole^{-1}.T^{-1}$ from water-methanol to pure water). It must also be noted that the increase of the activation entropy is even more important in the case of the diene **2β** than for methoxybutadiene. This could be related with the interaction of the glucose moiety with the water structure and could have some importance in connection with biological problems. In fact, it seems that glucose acts as a structure-making compound and enhances the hydrophobic effect. We have thus demonstrated, as previously postulated, that the rate enhancements in aqueous cycloadditions have an entropic origin.

Table IV. Second-order rate constants and activation parameters for the aqueous cycloadditions of $ROCH=CH-CH=CH_2$ with butenone

R	Solvent	$k.10^5$ ($M^{-1}.s^{-1}$) (25°C)	$\Delta H^{\dagger}$ ($kJ. mol^{-1}$)	$\Delta S^{\dagger}$ ($J.mol^{-1}.T^{-1}$)
β-D-glucosyl	H_2O	28.2	40.0±0.7	-178.8±2.4
id	H_2O -MeOH (50-50)	7.5	33.6±0.8	-211.1±2.6
CH_3	H_2O	114	38.9±0.7	-170.7±2.5
id	H_2O -MeOH (50-50)	23.4	35.4±1.0	-195.0±3.4

Conclusions.

We have shown that glyco-organic substrates react in aqueous cycloadditions to give goods yields of adducts through virtually pure *endo* transition states, more rapidly and at a much lower temperature, than the corresponding reactions in organic solvent. In the case of buta-1,3-dienyl glucosides, the facile removal of the sugar moiety by acidic or enzymatic hydrolysis yields highly functionalized chiral derivatives which can be further elaborated. In the case of 1,3-dienes having a stereogenic allylic hydroxy group and a second hydroxy group in β position, we postulate that the increase of facial selectivity in pure water compared with toluene could be a consequence of a difference of the hydrophobicity between the two faces of the dienes. Finally, the entropic origin of the rate enhancement has been clearly demonstrated.

Acknowledgments.

The authors thank the C.N.R.S. and the University Paris-Sud for financial supports.

Literature cited.

(1) David, S.; Eustache, J.; Lubineau, A. *C. R. Acad. Sc. Paris* **1973**, *276*, 1465.
(2) David, S.; Lubineau, A.; Vatele, J.M. *New J. Chem.* **1980**, *4(8/9)*, 547.
(3) Rideout, D.C.; Breslow, R. *J. Am. Chem. Soc.* **1980**, *102*, 7816.

(4) Grieco, P.A.; Garner, P.; He, Z. *Tetrahedron Lett.* **1983**, *24(18)*, 1897.
(5) Lubineau, A.; Queneau, Y. *Tetrahedron Lett.* **1985**, *26(22)*, 2653.
(6) Lubineau, A.; Queneau, Y. *J. Org. Chem.* **1987**, *52*, 1001.
(7) Lubineau, A.; Queneau, Y. *Tettrahedron* **1989**, *45(21)*, 6697.
(8) Klemer, A. *Chem. Ber.* **1963**, *96*, 634.
(9) Good, F.; Schuerch, C. *Carbohydr. Res.* **1984**, *125*, 165.
(10) Schmidt, O. T. *Methods in Carbohydrate Chemistry, vol 11*, **1963**, 318.
(11) Utamura, T.; Kuromatsu, K.; Suwa, K.; Koizumi, K.; Shingu, T. *Chem. Pharm. Bull.* **1986**, *34*, 2341.
(12) Klimko, V.T.; Scoldinov, A.P. *Zh. Obsh. Khim.* **1959**, *29*, 4027. *Chem. Abstr.* 54:208706b.
(13) Gupta, R.C.; Slawin, A.M.Z.; Stoodley, R.J.; Williams, D.J. *J. Chem. Soc., Chem. Comm.* **1986**, 1116.
(14) Brewster, J.H. *J. Am. Chem. Soc.* **1959**, *81*, 5483.
(15) Jurczak, J.; Pikul, S.; Bauer, T. *Tetrahedron* **1986**, *42*, 447.
(16) Lubineau, A.; Augé, J.; Lubin, N. *J. Chem. Soc., Perkin Trans I* **1990**, 3011.
(17) Frank, R.W.; Argade, S.; Subramanian, C.S.; Frechet, D.M. *Tetrahedron Lett.* **1984**, *26*, 3187.
(18) Kahn, S.D.; Hehre, W.J. *J. Am. Chem. Soc.* **1987**, *109*, 663.
(19) Kaila, N.; Frank, R.W.; Dannenberg, J.J. *J. Org. Chem.* **1989**, *54*, 4206.
(20) Tanford, C. *The Hydrophobic Effect* ; Wiley : New-York, 1980.
(21) Schneider, H.J.; Sangwan, N.K. *J. Chem. Soc., Chem. Comm.* **1986**, 1787.
(22) Lubineau, A. *J. Org. Chem.* **1986**, *51*, 2142.
(23) Lubineau, A.; Meyer, E. *Tetrahedron* **1988**, *44(19)*, 6065.
(24) Lubineau, A.; Augé, J.; Bellanger, N.; Caillebourdin, S. *Tetrahedron Lett.* **1990**, *31(29)*, 4147.
(25) Brandes, E.; Grieco, P.A.; Gajewski, J. *J. Org. Chem.* **1989**, *54*, 515.

RECEIVED December 2, 1991

Chapter 11

Stereoselectivity of 1,3-Dipolar Cycloaddition of Glycosyl Nitrones to *N*-Arylmaleimides

L. Fišera[1], U. A. R. Al-Timari[1,3], and P. Ertl[2]

[1]Department of Organic Chemistry, Slovak Technical University, CS–812 37 Bratislava, Czechoslovakia
[2]Chemical Institute, Comenius University, CS–842 15 Bratislava, Czechoslovakia

The cycloaddition of 3'-hydroxyglycosyl-*N*-methylnitrone 1 to *N*-arylmaleimides gave the syn isoxazolidines 6, whereas 3'-acetoxyglycosyl-*N*-methylnitrone 2 afforded the anti isoxazolidines 8. The formation of 6 was rationalized by an exo attack, stereoelectronically preferred through the hydrogen bond between the pentose hydroxyl group and one of the carbonyl group of *N*-arylmaleimide. The sterically preferred endo attack avoiding the repulsions between *N*-arylmaleimide and sugar moiety was proposed for addition of 2. Nitrones 13 derived from *D*-glucose oxime and benzaldehydes without employing any protection of hydroxyl group were isolated in a pure state. The addition of 13 to *N*-arylmaleimides gave predominantly the anti isoxazolidines 15 and was rationalized by Z/E isomerization of *N*-glycosylnitrones 13. AM1 calculations of the nitrones and MM2 calculations of adducts were performed.

Cyclic glycosides are important as enzyme inhibitors (*1*) and as chiral synthons, suitable for the synthesis of many natural products (*2*). These properties have stimulated interest in the synthesis of analogues. The 1,3-dipolar cycloaddition reaction has a nearly singular capability of establishing large numbers of stereochemical centers in one synthetic step (3). An impressive effort has been devoted to the synthetic application of the cycloaddition of nitrones to alkenes to give isoxazolidines (*3-6*). A large part of the research of stereocontrolled versions of 1,3-dipolar cycloaddition in the last few years has dealt with the influence exerted by a stereocenter located in either one of the two cycloaddends (*7*). In our efforts to utilize heterocyclic compounds as dipolarophile component in 1,3-dipolar cycloaddition reactions (*8*) we have recently demonstrated (Al-Timari U.A.R. and Fišera L. *Carbohydrate Research*, in press) that nitrile oxides and nitrones react with chiral sugar-derived alkenes to produce mainly anti-adducts with $\geq 95\ \%$ π-facial stereoselectivity. We have extended our studies to the preparation and cycloaddition chemistry of chiral nitrones.

In most of the known nitrones derived from carbohydrates, the dipole nitrogen is attached to the anomeric carbon. In the most of the chiral nitrones a fundamental role

[3]Current address: Basra University, Technical University, P.O. Box 272, Basra, Iraq

0097–6156/92/0494–0158$06.00/0

was played by *N*-sugar-derived nitrones (*9, 10*). Only scattered reports deal with chiral nitrones in which the sugar is attached at a different position (7). In this paper, we describe the preparation and 1,3-dipolar cycloaddition of a series of closely related *C*-sugar substituted nitrones and nitrones derived from *D*-glucose oxime to *N*-arylmaleimides. We also discuss the stereochemical aspects of these cycloadditions and show how AM1 calculations can be used and refined to predict the stereochemical outcome of such cycloadditions.

Preparation of C-Glycosyl Nitrones

A series of novel 3'-hydroxy-*N*-methyl- **1**, 3'-acetoxy-*N*-methyl- **2** and 3'-acetoxy-*N*-phenylnitrones **3** was prepared, isolated, and treated with *N*-arylmaleimides. Preparation of the nitrones was accomplished from the corresponding aldehydes **4** and **5** (*11*), which upon treatment with *N*-methylhydroxylamine and *N*-phenylhydroxylamine were converted to the Z-nitrones **1, 2** and **3**, respectively. The geometry of nitrones **1** and **2** was verified by an NOE experiment which showed an enhancement of the *N*-methyl signal upon irradiation of the azomethine hydrogen as well as an enhancement of the azomethine proton signal upon irradiation of the *N*-methyl group. The coupling constant $J_{4',5}$ = 5.1 Hz is indicative of a gauche H-4' and H-5 relationship.

1: R=H, Y=N(O)Me
2: R=Ac, Y=N(O)Me
3: R=Ac, Y=N(O)Ph
4: R=H, Y=O
5: R=Ac, Y=O

1,3-Dipolar Cycloadditions of Nitrones 1 - 3 to *N*-Arylmaleimides

It was found that 3'-hydroxy-*N*-methylnitrone **1** reacted with *N*-arylmaleimides in toluene at 110°C to give the syn-isoxazolidines **6** (Scheme I, H-3,H-3a syn-relationship). In contrast, the 3'-acetoxy-*N*-methyl- and 3'-acetoxy-*N*-phenylnitrones **2** and **3** gave the anti-isoxazolidines **8** respectively (H-3, H-3a anti-relationship). It would appear that the introduction of an acetoxy groups into a C-3' carbon of nitrone can significantly influence its stereochemical behaviour.

Proton NMR analysis of isoxazolidines **8** revealed that each diastereomer has an anti relationship between H-3 and H-3a. In **8a**, for example, the signal for the H-3a proton appears as a doublet at δ 3.42 with a coupling constant of 7.8 Hz. This feature of the

The cycloaddition results in the generation of three new asymmetric centres; however, the condensed adducts are constrained to a cis-arrangement of H-3a and H-6a bridgehead protons, therefore, four diastereomeric cycloadducts were possible, (Scheme II)

Stereochemical assignments of H-3, H-3a and H-6a atoms were made to the condensed isoxazolidines on the basis of spectroscopic data, in particular using the $J_{3\text{-}3a}$

Scheme I

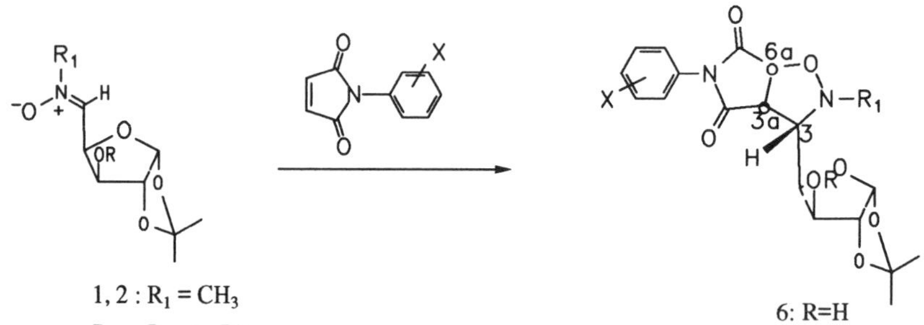

$1,2 : R_1 = CH_3$

$3 : R_1 = Ph$

6: R=H

9: R=Ac

a) X= H

b) X= 2,6–diCH_3

c) X= 2–CH_3,6–C_2H_5

d) X= 2,6–diC_2H_5

e) X= 2,5–diCl

f) X= 4–Cl

g) X= 4–F

h) X= 4–Br

i) X= 3,4–diCl

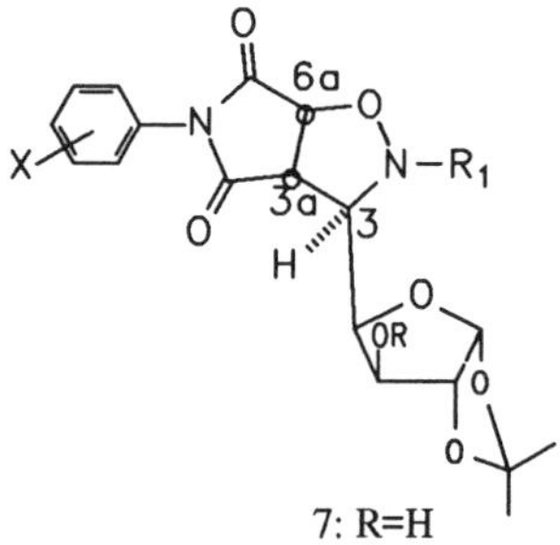

7: R=H

8: R=Ac

and $J_{3a\text{-}6a}$ coupling constants. The ring junction between the two rings was always cis, as indicated by coupling constants and an examination of molecular models. Moreover, the 1,3-dipolar cycloaddition reactions of nitrones to alkenes are known to proceed with cis-stereospecificity (*3*). For instance, in the compound **6a** the coupling constant for the cis ring junction protons H-6a and H-3a is equal to 7.2 Hz, and in **8a**. 7.8 Hz. These values indicate a nearly eclipsed dihedral angle between H-3a and H-6a.

NMR spectrum is uniquely diagnostic for the H-3, H-3a anti relationship (*12*). Proton H-3 is coupled solely to H-4' with the coupling constant $J_{3\text{-}4'}$ = 8.4 Hz, indicative of nearly eclipsed conformation of **8a** between H-3 and H-4'. Moreover, ^1H NMR data for the **8** enable almost complete configurational assignments to be made, e.g. for the **8b** irradiation of H-3' caused NOE's for both H-4'and H-3a, which suggested that these three protons were all on the same side. Irradiation of H-3a results in signal enhancement of H-6a and of H-3' of the saccharide unit, consistent with the cis-configuration for H-3a and H-6a.

In **8** the 0-1 Hz coupling constant between bridgehead H-3a and isoxazolidine H-3 is consistent only with anti stereochemistry, since in the syn isomer (**9**) the two hydrogens would be nearly eclipsed, and give rise a much larger coupling constant.

Indeed, the adduct **6**, from the cycloaddition of 3'-hydroxy-*N*-methylnitrone **1a**, showed $J_{3\text{-}3a} = 7.2$ Hz, which is in the range expected of a H-3, H-3a syn relationship. The further support for this syn arrangement is that, the signal for the H-3a proton appears as a doublet of doublets, and H-3 appears as a multiplet.

Scheme II

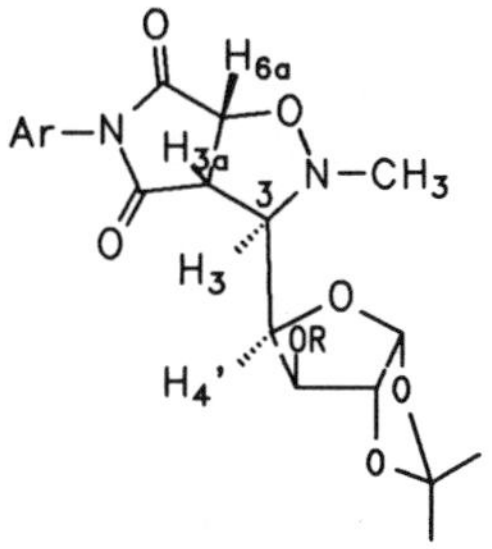

anti – syn 7, 8

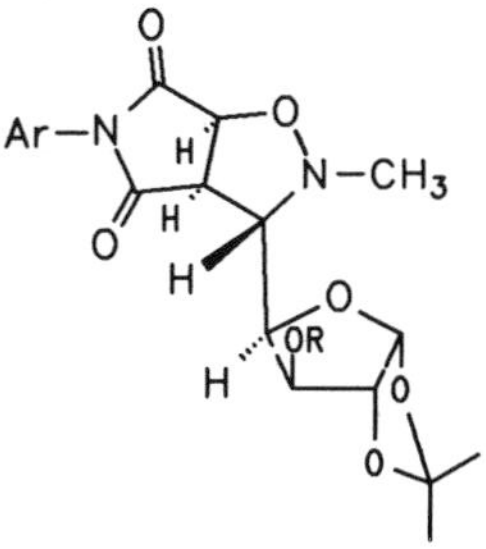

anti – anti 7, 8

H-3a,H-6a = 7.2 Hz
H-3, H-4' = 8.4 Hz

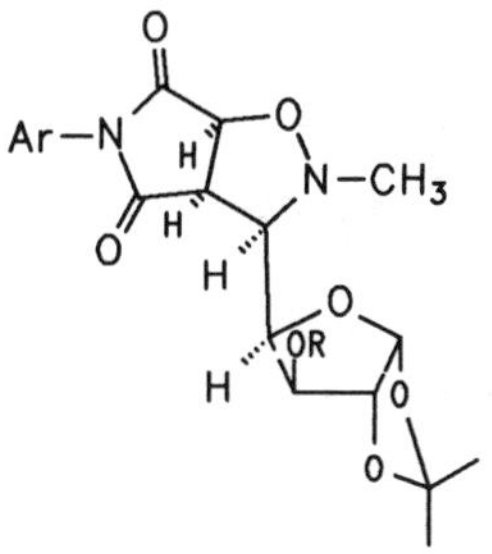

syn – syn 6, 9

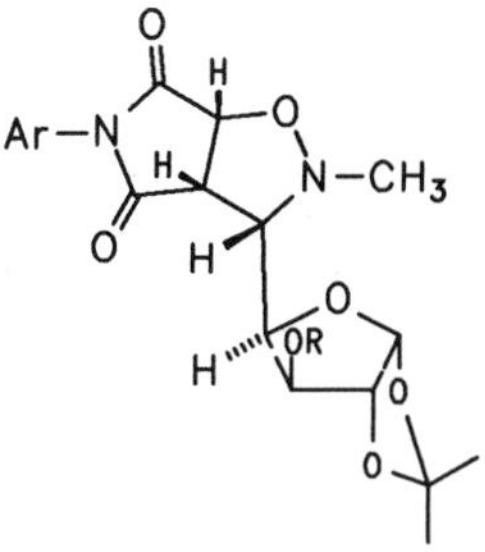

syn – anti 6, 9

H-3a, H-6a = 7.2 Hz
H-3, H-3a = 7.2 Hz

It was not possible from the spectroscopic data available to decide if the anti isoxazolidines **8** obtained from the 3'-acetoxy-*N*-methyl- (or phenyl) nitrone and *N*-arylmaleimide corresponded to anti-syn isomer **8** or to anti-anti isomer **8** and if the syn isoxazolidines **6**, which have been isolated from the cycloaddition of the 3'-hydroxy-*N*-methylnitrone to *N*-arylmaleimide corresponded to syn-syn isomer **6** or syn-anti isomer **6**, respectively .

Stereoselectivity of 1,3-Dipolar Cycloaddition. The stereoselectivity of the intermolecular cycloaddition of an acyclic nitrone to an alkene is difficult to predict, and would appear to be susceptible to minor structural changes in either component (*13*). The chiral 2,2-dimethyl-1,3-dioxolan-4-yl nitrone showed only modest diastereoface selectivity in its addition to methyl crotonate (*14*). However, the more hindered tetramethyl-1,3-dioxolan-4-yl nitrone was more selective.

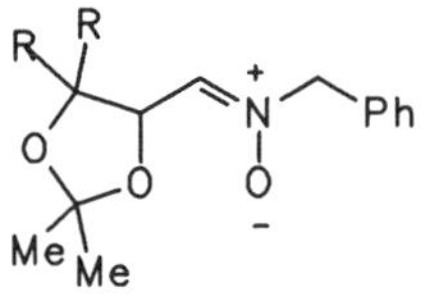

R = H, Me

As anticipated from earlier studies (*12*), the anti C-3, C-4' cycloadducts were produced stereoselectively or predominated. Deshong and coworkers showed, that the cycloaddition of α,β-dialkoxy substituted nitrones with vinylene carbonate displayed moderate selectivity for the diastereomer having a C-3, C-4' anti relationship (*4*).

Dipolar cycloaddition of α-alkoxy substituted nitrones had been shown to occur preferentially via Felkin-Anh (*15-17*) transition states in which the developing carbon-carbon bond avoids steric interaction with the bulky group (*4*). Accordingly, in our case the conformations A and B can be invoked (Figure 1).

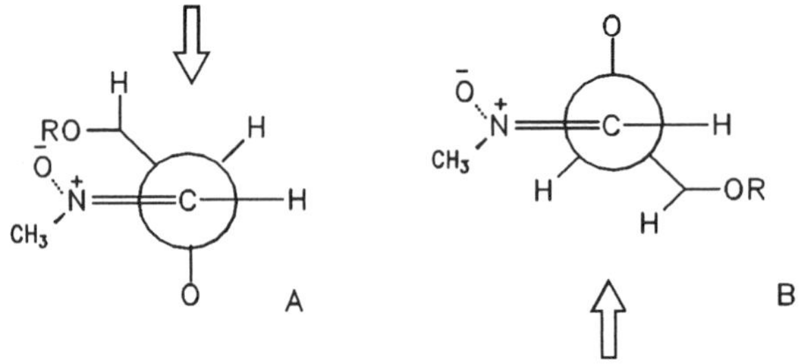

A B

exo attack ► syn – syn exo attack ► syn – anti
endo attack ► anti – syn endo attack ► anti – anti

Figure 1. Conformations of α-alkoxy substituted nitrones.

All the nitrones used in this study were isolated before cycloaddition, and were shown by ^{1}H NMR NOE data to be the Z-isomers implicit in formulas **1, 2** and **3**. The isomeric E-nitrones could not be detected, but as it has been postulated that isomerization of Z-nitrones to the more reactive E-nitrones can precede cycloaddition (*18*). It is not possible to exclude that either or both Z- and E-nitrones are involved in cycloadditions. The anti-isoxazolidines **7** and **8** can arise from cycloaddition of Z-nitrones through endo transition states, or from reaction of the E-nitrone in an exo-mode. Conversely the syn-isoxazolidines **6** and **9** could be formed by the Z-nitrone reacting in the exo-fashion, or the E-nitrone in an endo-mode. (Scheme III).

Scheme III

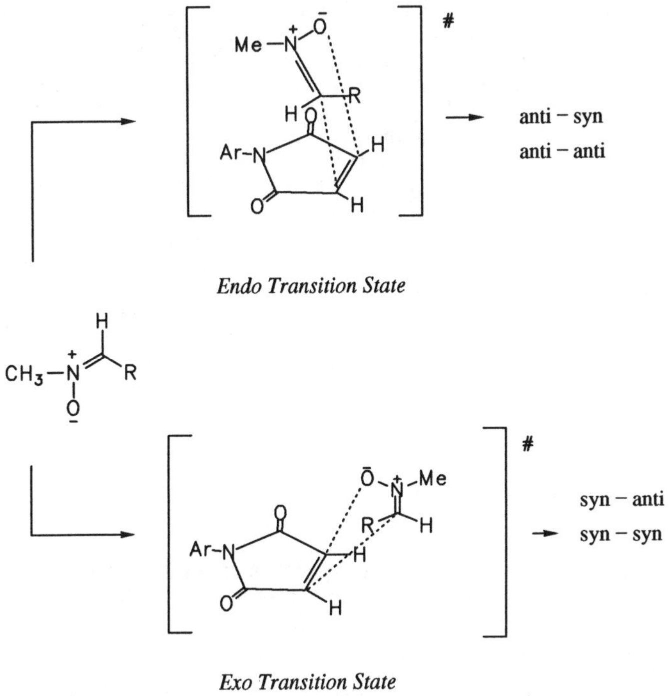

anti – syn
anti – anti

Endo Transition State

syn – anti
syn – syn

Exo Transition State

Accordingly, the cycloadditions via transition state **A** (Figure 1) would afford syn-syn products through exo-attack and anti-syn products through endo-attack. Conversely, cycloaddition via transition state **B** would furnish syn-anti cycloadducts through exo-attack and anti-anti adducts through endo-attack.

In order to rationalize the above cycloadditions, we have carried out quantum mechanical calculations. The relative stabilities of products with anti-syn, anti-anti, syn-anti and syn-syn configurations have been assessed by semiempirical quantum mechanical calculations (AM1 method) (*19*). Geometries of both nitrones and cycloadducts were totally optimized. Subsequent AM1 calculations showed the Z-nitrone **1** to be more stable by 14.8 kJ/mol than the corresponding E-isomer, a fact that can be accounted for mainly by steric considerations. Inspection of frontier orbital energies shows that the interaction of **1** with *N*-phenylmaleimide is governed by the HOMO dipole. The calculated relative energies in kJ/mol are expressed as energy differences, the energy of the most stable structure being the reference.

anti-syn = 0.0 anti - anti = 1.5 syn - anti = 1.8 syn - syn = 34.1

Thus, the 3'-hydroxy-*N*-methylnitrone **1** gave the syn-anti cycloadducts **6** via exo attack (conformation **B**), since the formation of syn-syn adducts is possible from the AM1 calculations to neglect. Molecular models suggest that attack via mode **B** is the least sterically demanding, and may explain the selective formation of the syn-anti adducts in the case of the hydroxynitrones, and the anti-anti adducts in the case of acetoxynitrones. For conformation **A**, severe steric interactions are generated between the incoming *N*-arylmaleimide and sugar moiety, and unfavourable interactions between the nitrone oxygen and the substituent on the C-3' carbon are present. We propose that the aforementioned nitrones undergo cycloaddition via conformation **B**, favoured on steric grounds. Endo attack is sterically preferred, since the repulsions between *N*-arylmaleimide and sugar moiety are avoided for the 3'-acetoxynitrones, and exo attack is stereoelectronically preferred because of the hydrogen bond between the pentose hydroxyl group and one of the carbonyl groups of the *N*-arylmaleinimide.

NMR analysis of the crude mixture showed the presence less than 10% the second anti - syn isomer in the cycloadditions of 3'-acetoxynitrones, and the syn-syn isomer in the case of 3'-hydroxynitrone. These compounds was not possible to separate from major products. Cycloaddition of nitrones **1** and **2** and bis-maleimide **10** proceeded analogously and resulted in the stereoselective formation of syn-adducts **11** with the nitrone **1** and anti-adducts **12** for the nitrone **2** (Scheme IV).

Thus the cycloaddition of the *C*-glycosylnitrones to *N*-arylmaleimides would appear to proceed with useful stereoselectivity, but the nature of the stereoselectivity is dependent upon the precise functionality present in the nitrone. We have demonstrated that small structural change in the nitrone result in significant changes in the steroselectivity of cycloaddition.

Preparation of D-Glucose-derived Nitrones.

Among the chiral nitrones possessing stereocenters in the substituent at nitrogen, a fundamental role was played by sugar-derived nitrones (*7*). In a series of papers by Vasella and co-workers (*9*), it was found that mannose- and ribose-derived nitrones afforded up to 95% stereoselectivity at C-5 in cycloadditions to methyl methacrylate (*20*). Stereocontrol at the C-3 position of the isoxazolidine ring is much lower and depends on the nature of the substituent in the aforementioned nitrones (*21*). Recently, Sandhu and coworkers described the succesful *in situ* preparation of nitrone **13** derived from *D*-glucose oxime and benzaldehyde, without protecting hydroxyl groups (*22*). It has been the first example of this type of dipole, but nitrone **13** was not isolated, and was reacted *in situ* with dipolarophiles such as acrylonitrile, styrene and *N*-phenylmaleimide. The stereochemistry of the so-prepared isoxazolidines has been not established (*22*).

We have prepared and isolated in a pure state nitrones **13a-g** derived from *D*-glucose oxime and X-substituted benzaldehydes (where X is 4-Cl, 2,4-diCl, 2,6-diCl, 2,4,6-triCH_3, 4-NO_2, 3-NO_2) and treated them with *N*-arylmaleimides. Preparation of the nitrones was accomplished from the corresponding benzaldehydes, which were converted to the Z-nitrones **13**, by treatment with *D*-glucose oxime **14** in dry ethanol at room temperature (Scheme V). The nitrones, **13** possessing an electron-withdrawing substituent on the benzene ring, are stable crystalline compounds, but

Scheme IV

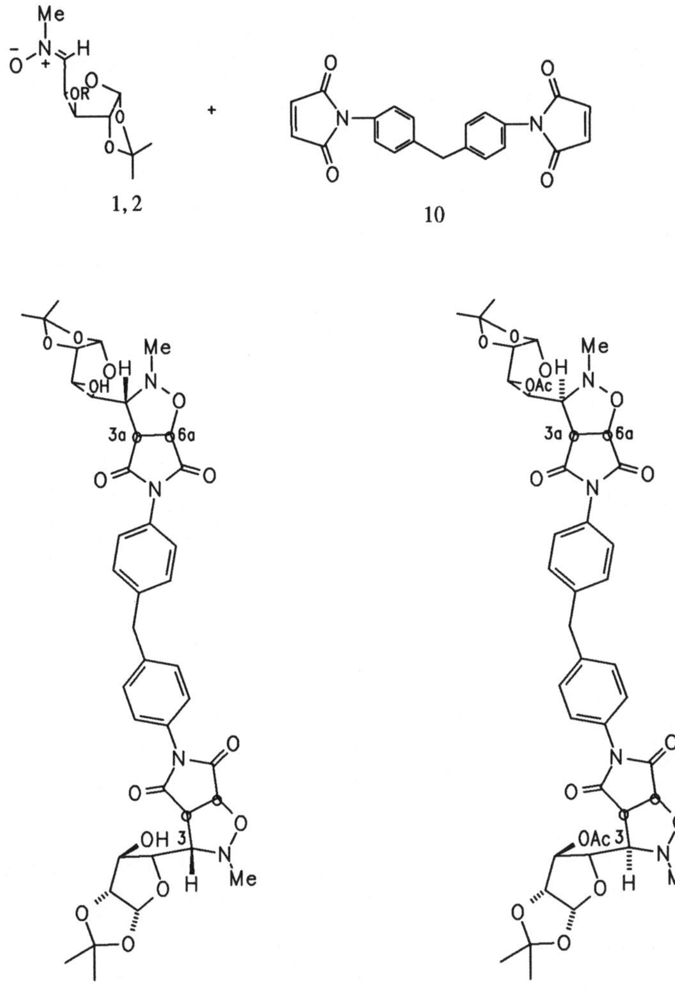

1,2

10

11

12

they are not stable to treatment with water. In wet ethanol, hydrolysis occurs to give the starting benzaldehydes and *D*-glucose oxime. The nitrones having an electron-donating group such as methyl or methoxy can be prepared only in situ. On the other hand the o,o-disubstituted nitrones are unusually stable. All nitrones **13a-g** are diastereomerically pure.

Scheme V

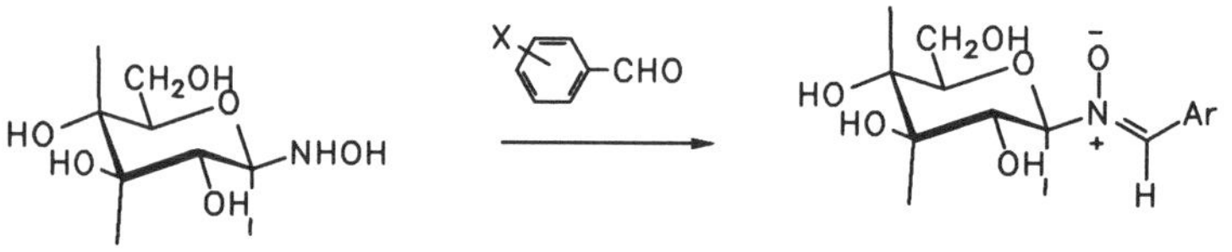

14

13

a, X=4–Cl

b, X=2,4–diCl

c, X=2,6–diCl

d, X-4–CH_3

e, X=2,4,6–triCH_3

f, X=4–NO_2

g, X=3–NO_2

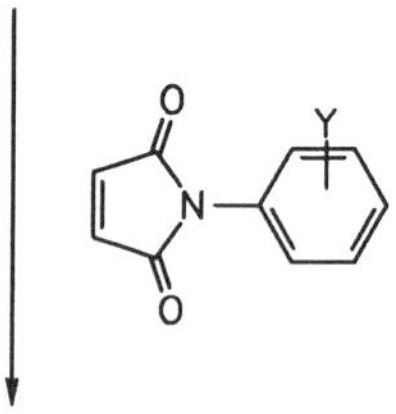

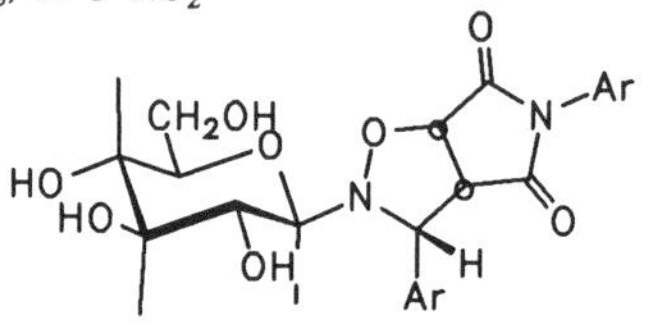

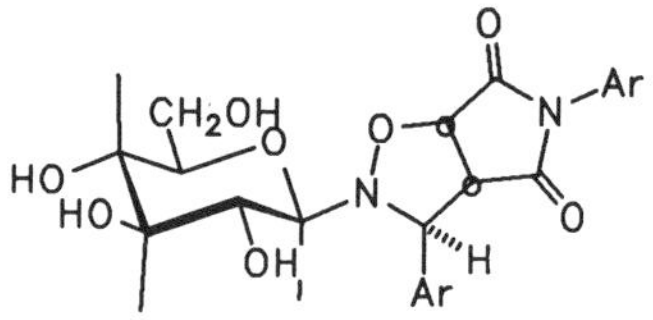

16

15

a, X=4–Cl, Y=3,5–diCl

b, X=4–Cl, Y=2,6–diCH_3

c, X=4–NO_2, Y=3,5–diCl

d, X=4–CH_3, Y=3,5–diCl

e, X=4–Cl, Y=4–iPr

^{1}H NMR spectra of nitrones **13**, in deuterated dimethyl sulphoxide were analysable, including the assignment of C-OH signals. Spectral assignments were made with the aid of exchange in deuterium oxide, double quantum filtered ^{1}H-^{1}H correlation spectroscopy (DQF ^{1}H-^{1}H COSY) and ^{1}H-^{13}C COSY. Some ambiguity remained in assignment of proton signals of the sugar moiety because of overlap in the proton spectra at δ 3.10-3.75. For example, the signal at δ 8.09 for the 4-chloroderivative **13a** was assigned to H-1 nitrone proton. The ^{1}H-NMR spectrum of **13a** posseses significant doublets at δ 5.48 (J=5.4 Hz), 5.19 (J=5.1 Hz), 5.10 (J=5.1 Hz), and a triplet at δ 4.73, which were assigned to OH groups. The proposed structure of **13a** is further supported by presence of doublets at δ 99.62 (C-1'), 79.98, 76.88, 70.01 and 69.55, as well as a triplet at δ 60.93. An indication for the β-pyranose form of nitrone **13a** is given by the presence of the doublet at δ 4.87 with the coupling constant J = 8.4 Hz for the H-1' proton. This is also supported by the fact that the starting *D*-glucose oxime **14** exists in the cyclic β-pyranose in the solid state and isomerises to a mixture of β-pyranose (23%), α-pyranose (7%), anti-(Z-)(13.5%) and syn-(E-)(56.5%) forms in aqueous solution (*23*). The exclusive formation of β-cyclic form is consistent with steric considerations. We suppose that the nitrones **13** have the Z-configuration, since there is strong evidence that nitrones derived from aromatic aldehydes possess a configuration in which the *C*-aryl and *N*-alkyl group are in a trans relationship (*24*). This consideration is in accord with Vasella's results which indicated that *N*-glycosylaldehydonitrones possess a Z-configuration (*20-21*).

It is a well known practice to derivatize polyhydroxyl compounds, prior to measuring their MS spectra, as more volatile derivatives; however, we measured the mass spectra of the native polyhydroxyl compounds. The EI mass spectra of compounds **13a, 13b, 13c, 13e** showed molecular ion peaks of very low intensity (less than 1%). These compounds also demonstrated a loss of 16 atomic units (atomic oxygen), typical of spectra of nitrones. The typical degradation process in this type of compound was the loss of the sugar component from the molecular ion, acompanied by a hydrogen shift, to give fragment ions fo the general formula

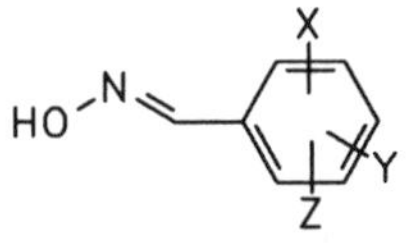

I

The fragmentation of the aforementioned ion then proceeds by the loss of fragments such as OH., H_2O, HCN, as well as CH_3., CO., Cl., depending on the substituents present.

1,3-Dipolar Cycloadditions of Nitrone 13 to *N*-Arylmaleimides.

1,3-Dipolar cycloaddition of *N*-glycosylnitrones **13** and *N*-arylmaleimides in ethanol at 78°C afforded the isoxazolidines **15** and **16** as an inseparable mixture of diastereoisomers, with the preponderance of the anti isomer **15**. The ratio of anti to syn diastereoisomers was: **15a** : **16a** (77:23); **15c** : **16c** (70:30); **15d** : **16d** (95:5); **15e** : **16e** (75:25), as determined by integration of the H-3a signals in the ^1H NMR spectra.

In contrast to the above examples, the 1,3-dipolar cycloaddition to *N*-(2,6-dimethylphenyl) maleimide of the *N*-glycosylnitrone **13** gave predominantly the syn-isoxazolidine **16b**, with a diastereomeric excess of more than 90%. The crude residue after cycloadditions was chromatographed, and the corresponding major products could be obtained more then 90% purity with a small amount of the minor component. In the case of o,o-disubstituted *C*-(2,4,6-trimethylphenyl)-**13e** and *C*-2,6-dichlorophenyl-*N*-glycosyl-nitrones **13c**, the cycloaddition reaction with *N*-arylmaleimides did not take place; only the unreacted starting compounds were isolated.

Three new asymmetric centers C-6a,C-3a and C-3 were generated in the cycloaddition of *N*-glycosylnitrones **13** and *N*-arylmaleimides, since the condensed adducts possessing a cis-relationship of H-3a and H-6a bridgehead protons, therefore, four diastereomeric products, syn-anti **15**, anti-anti **15**, anti-syn **16** and syn-syn **16** could be formed. The first prefix anti or syn showed a relationship between H-1' and H-3 atoms and the second a relationship between H-3 and H-3a atoms (Scheme VI).

Scheme VI

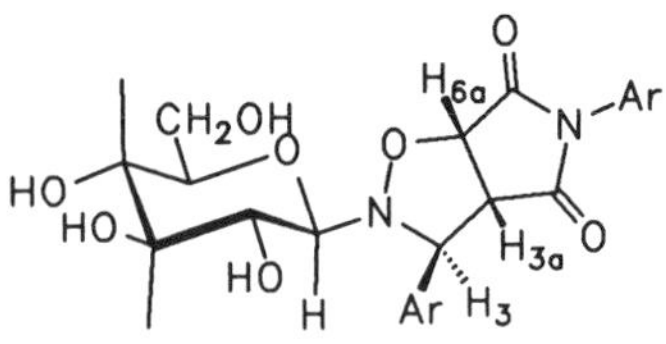

syn – anti 15

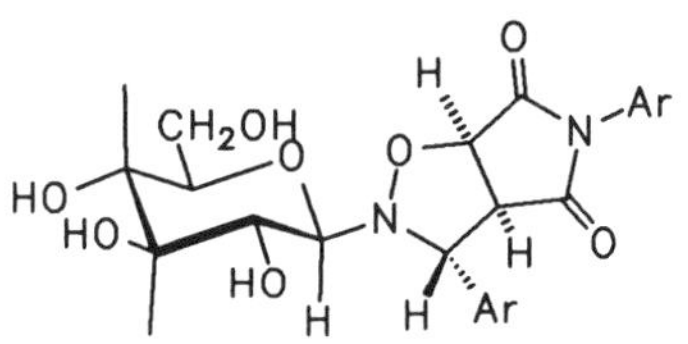

anti – anti 15

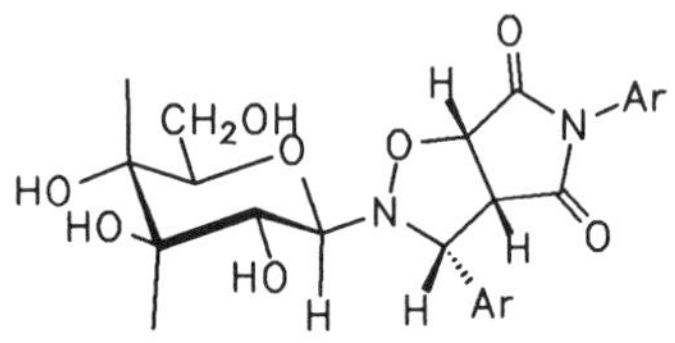

anti – syn 16

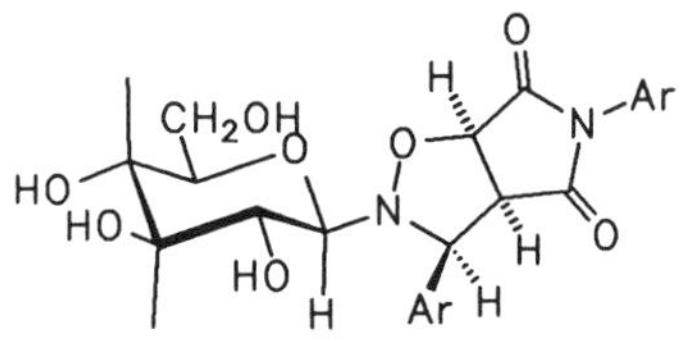

syn – syn 16

The distinction between the arrangements of H-3, H-3a and H-6a atoms was done by spectroscopic data, in particular using the $J_{3\text{-}3a}$ and $J_{3a\text{-}6a}$ coupling constants. Proton NMR analysis of major isoxazolidines **15** revealed that each diastereomer had an H-3, H-3a anti relationship. In **15a**, for example, the signals for the H-6a and H-3a appeared as doublets at δ 5.15 and 3.90, respectively, with a coupling constant of $J_{3a\text{-}6a}$ 8.4 Hz from coupling solely to H-6a. Proton H-1' at δ 5.22 is coupled solely to H-2' with the coupling constant $J_{1'\text{-}2'}$= 8.4 Hz, indicative of the β-cyclic form of the sugar moiety. On the other hand, the major adduct **16b**, from the cycloaddition of *N*-(2,6-dimethylphenyl)maleimide, as well as the minor products **16** from the cycloaddition of other *N*-arylmaleimides, showed the presence of doublets for H-6a protons and doublets of doublets for H-3a protons.

Stereoselectivity of 1,3-Dipolar Cycloaddition. The anti-isoxazolidines **15** arise from cycloaddition of Z-nitrone through endo transition state, or the E-nitrone in an exo-mode. Conversely the syn-isoxazolidines **16** could be formed by the Z-nitrone reacting in the exo-fashion or the E-nitrone in an endo-mode (Scheme VII).

Both anti-**15** and syn-**16** adducts obtained in cycloadditions using *N*-arylmaleimide as the dipolarophile must arise from the exo transition state. The E-isomer of the nitrone **13** yields the anti adduct **15**, while the Z-isomer yields the syn adduct **16**. Endo transition states would suffer from unfavourable steric interactions between the *N*-arylmaleimide moiety in the incoming dipolarophile and the *N*-glycosyl moiety of

Scheme VII

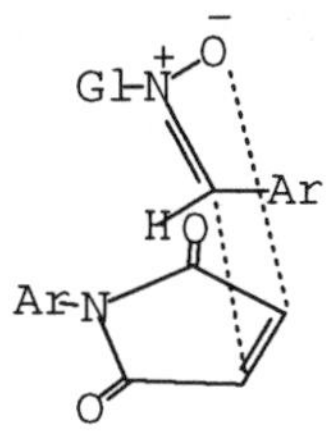

A

endo TS → anti 15

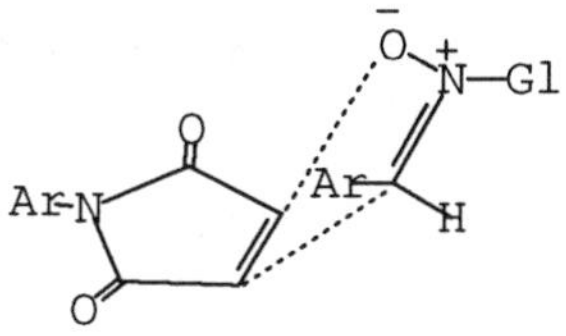

B

exo TS → syn 16

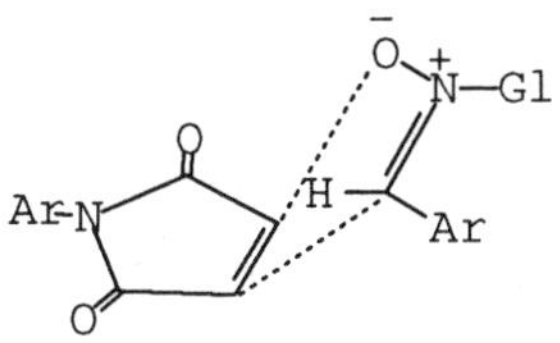

C

exo TS → anti 15

the nitrone. Nitrone isomerization (Z→E) was invoked by Vasella to account for the diastereoselectivity of the 1,3-dipolar cycloaddition of *N*-sugar derived nitrones with ethylene (*21, 25*). The formation of **15** via transition state **C** in comparison with the formation of **16** through transition state **B** is favourable for steric reasons. The van der Waals nonbonding interaction between the two aryl groups in transition state **B** should be larger than in transition state **C**, involving the E-form of nitrone **13**.

It was not possible from the spectroscopic data available to establish the syn- or anti- relationship between H-1' and H-3. In a series of papers by Vasella and co-workers, (*20, 21*) it was found that the 1,3-dipolar cycloaddition of *N*-glycosylnitrones to methyl methacrylate gave *N*-glycosylisoxazolidines with a high degree of diastereoselectivity (diastereomeric excess of over 90%). The observed diastereoselectivity has been rationalized on the basis of a stereoelectronic effect in combination with steric effects. AM1 calculations showed, that the steric effects, which determine both the relative population of the relevant conformers and the direction of attack of the 1,3-dipolarophile, appear to be more important than the difference of the stereoelectronic effects in the antiperiplanar *vs.* synperiplanar orbital arrangement (*9*). According to these results, the O-endo conformer of nitrone **13** should be preferred in the ground state, since the steric interaction between the E-nitrone *C*-substituents and the glycosyl moiety destabilize the O-exo conformation. Moreover, in Z,O-endo conformation, there is an intramolecular hydrogen bond between the C-2' hydroxyl group and the oxygen of the nitrone. Its length was found to be 2.12 Å. (Scheme VIII). The hydrogen bond is also possible in E,O-endo conformation, however, in this case strong steric repulsions between *C*-phenyl and *N*-moieties are present. The calculated relative energies (AM1) in kJ/mol are expressed as energy differences, the energy of the most stable structure being the reference.

Z,O - exo = 5.4 Z,O - endo = 0.0 E,O - exo = 8.9 E,O - endo = 20.4

Scheme VIII

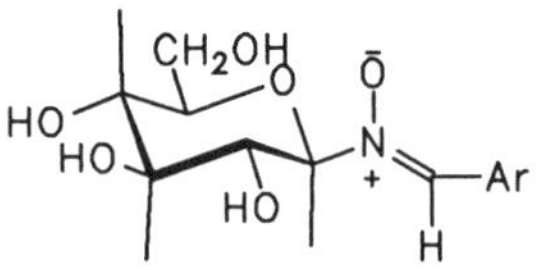

Z,-O-exo

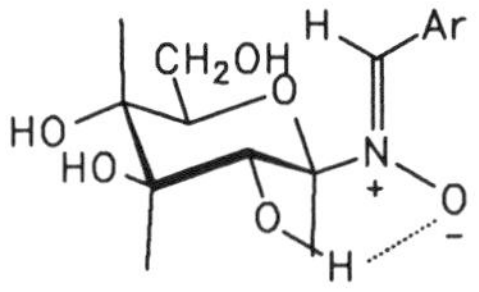

Z,-O-endo

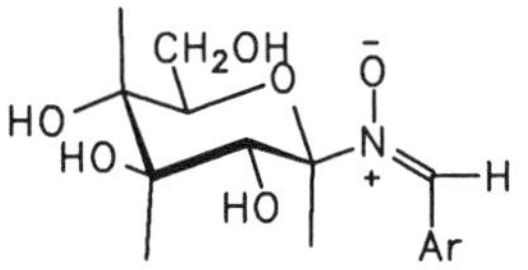

E,-O-exo

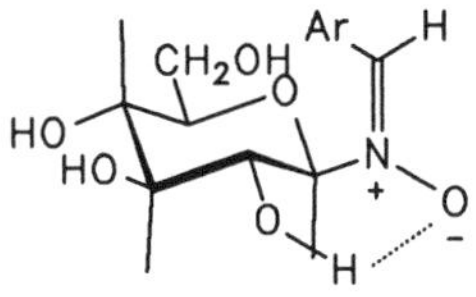

E,-O-endo

Dipolar cycloaddition of chiral nitrones was shown to occur preferentially via a Felkin-Anh (*15-17*) transition state in which the developing carbon-carbon bond avoided steric interaction with the bulky group (4). Both conformations **D** (the ring oxygen atom is perpendicular to the plane of the nitrogen-carbon double bond) and **E** (the C-2' carbon atom is perpendicular) can be stabilized by the presence of an intramolecular hydrogen bond, and produce the syn-anti product **15** by an anti attack (Figure 2), and the anti-anti product **15** by a syn attack. Inspection of frontier orbital energies shows, that the interaction of **13** with *N*-phenylmaleimide is governed by the HOMO dipole. The calculated relative energies of the possible four cycloadducts in kJ/mol are expressed as energy differences, the energy of the most stable structure being the reference.

syn - anti = 2.1 anti -anti = 1.0 anti - syn = 0.0 syn - syn = 11.7

From the hypothesis that the E-configuration of the nitrone **13** reacts via its *O*-exo-conformer, and that an anti-attack in exo transition state is preferred, the formation of anti-anti **15** adducts may be expected.

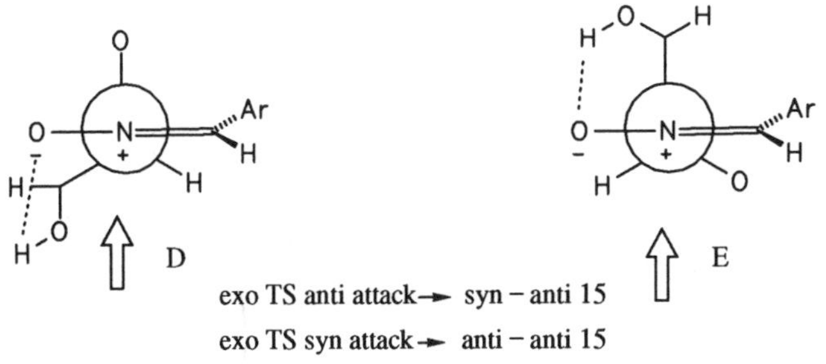

D E

exo TS anti attack → syn – anti 15

exo TS syn attack → anti – anti 15

Figure 2. Nitrone-conformers stabilized by hydrogen bonding.

Literature Cited

1. Buchanan, J. G. *Fortschr. Chem. Org. Naturst.* **1983**, 44, 243.
2. Hanessian, S. *Total Synthesis of Natural Products : The Chiron Approach*; Pergamon : Oxford, **1983**.
3. Tufariello, J. J. In *1,3-Dipolar Cycloaddition Chemistry* ; Padwa, A.; Wiley : New York, **1984**, Vol. 2; 83-168.
4. DeShong, P.; Li, W.; Kennington, Jr. J. W.; Ammon, H.L.; Leginus, J. M. *J. Org. Chem.* **1991**, 56, 1364.
5. Annunziata, R.; Cinquini, M.; Cozzi, F.; Giaroni, P.; Raimondi, L. *Tetrahedron Lett.* **1991**, 32, 1659.
6. Ito, M.; Kibayashi, C. *Tetrahedron Lett.* **1990**, 31, 5065.
7. Annunziata, R.; Cinquini, M.; Cozzi, F.; Raimondi, L. *Gazz. Chim. Ital.* **1989**, 119, 253.
8. Oravec, P.; Fišera, L. *Monatsh. Chem.* **1991**, 122, 165 and references therein.
9. Huber, R.; Vasella, A. *Tetrahedron* **1990**, 46, 33.
10. Iida, H.; Kasahara, K.; Kibayashi, C. *J. Am. Chem. Soc.* **1986**, 108, 4647.
11. Tronchet, J. M. J.; Mihaly, M. *Helv. Chim. Acta* **1972**, 55, 1266.
12. DeShong, P.; Dicken, C. M.; Leginus, J. M.; Whittle, R. R. *J. Am. Chem. Soc.* **1984**, 106, 5598.
13. DeShong, P.; Dicken, C.M.; Staib, R. R.; Freyer, A.J.; Weinreb, S. M. *J. Org. Chem.* **1982**, 47, 4397.
14. Fray, M. J.; Jones, R. H.; Thomas, E.J. *J. Chem. Soc. Perkin Trans* 1, **1985**, 2753.
15. Anh, N. T.; Eisenstein, O. *Nouv. J. Chem.* **1977**, 1, 61.
16. Bürgi, H. B., Dunitz, J.D.; Lehn, J.M.; Wipf, G. *Tetrahedron* **1974**, 30, 1563.
17. Houk, K. N.; Moses, S. R.; Wu, Y.-D.; Rondan, N. G.; Jäger, V.; Shohe, R.; Fronczek, F. R. *J. Am. Chem. Soc.* **1984**, 106, 3880
18. Kametani, T.; Huang, S.-P.; Nakayama, A.; Honda, T. *J. Org. Chem.* **1982**,47, 2328.
19. Dewar, M. J. S.; Zoebisch, E. G.; Healy, E. F.; Stewart, J. J. P. *J. Am. Chem. Soc.* **1985**, 107, 3902.
20. Vasella, A. *Helv. Chim. Acta* 1977, 60, 426, 1273.
21. Vasella, A.; Voeffray, R. *J. Chem. Soc. Chem. Commun.* **1981**, 97.
22. Borthakur, D. R.; Prajapati, D.; Sandhu, T. S. *Ind. J. Chem.* **1988**, 27B, 724.
23. Finch, P.; Merchant, Z. *J. Chem. Soc. Perkin Trans.* 1, **1975**, 1682.
24. Folting, K.; Lipscomb, W. N.; Jerslev, B. *Acta Chem. Scand.* **1963**, 17, 2138.
25. Huber, R.; Knierzinger, A.; Obrecht, J. P.; Vasella, A. *Helv. Chim. Acta* **1985**, 68,1730.

RECEIVED December 2, 1991

INDEXES

Author Index

Affiliation Index

Subject Index

A

B

C

Production: Donna Lucas
Indexing: Deborah H. Steiner
Acquisition: Barbara C. Tansill
Cover design: Amy Meyer Phifer

Printed and bound by Maple Press, York, PA

Other ACS Books

Chemical Structure Software for Personal Computers
Edited by Daniel E. Meyer, Wendy A. Warr, and Richard A. Love
ACS Professional Reference Book; 107 pp;
clothbound, ISBN 0–8412–1538–3; paperback, ISBN 0–8412–1539–1

Personal Computers for Scientists: A Byte at a Time
By Glenn I. Ouchi
276 pp; clothbound, ISBN 0–8412–1000–4; paperback, ISBN 0–8412–1001–2

Biotechnology and Materials Science: Chemistry for the Future
Edited by Mary L. Good
160 pp; clothbound, ISBN 0–8412–1472–7; paperback, ISBN 0–8412–1473–5

Polymeric Materials: Chemistry for the Future
By Joseph Alper and Gordon L. Nelson
110 pp; clothbound, ISBN 0–8412–1622–3; paperback, ISBN 0–8412–1613–4

The Language of Biotechnology: A Dictionary of Terms
By John M. Walker and Michael Cox
ACS Professional Reference Book; 256 pp;
clothbound, ISBN 0–8412–1489–1; paperback, ISBN 0–8412–1490–5

Cancer: The Outlaw Cell, Second Edition
Edited by Richard E. LaFond
274 pp; clothbound, ISBN 0–8412–1419–0; paperback, ISBN 0–8412–1420–4

Practical Statistics for the Physical Sciences
By Larry L. Havlicek
ACS Professional Reference Book; 198 pp; clothbound; ISBN 0–8412–1453–0

The Basics of Technical Communicating
By B. Edward Cain
ACS Professional Reference Book; 198 pp;
clothbound, ISBN 0–8412–1451–4; paperback, ISBN 0–8412–1452–2

The ACS Style Guide: A Manual for Authors and Editors
Edited by Janet S. Dodd
264 pp; clothbound, ISBN 0–8412–0917–0; paperback, ISBN 0–8412–0943–X

Chemistry and Crime: From Sherlock Holmes to Today's Courtroom
Edited by Samuel M. Gerber
135 pp; clothbound, ISBN 0–8412–0784–4; paperback, ISBN 0–8412–0785–2